# 国际空间科学小卫星发展路线图

## SMALL SATELLITES FOR SPACE SCIENCE

### A COSPAR SCIENTIFIC ROADMAP

（美）罗宾·米兰 （瑞士）鲁道夫·施泰格 等 著

郭世杰 许永建 译 李 靖 审校

U0190027

中国科学技术大学出版社

安徽省版权局著作权合同登记号：第 12201970 号

**图书在版编目(CIP)数据**

国际空间科学小卫星发展路线图：汉英对照/(美)罗宾·米兰(Robyn M. Millan)，(瑞士)鲁道夫·施泰格(Rudolf von Steiger)等著；郭世杰，许永建译.—合肥：中国科学技术大学出版社，2020.11

ISBN 978-7-312-05019-0

Ⅰ.国…　Ⅱ.①罗…②鲁…③郭…④许…　Ⅲ.科学卫星—小型卫星—研究—汉、英　Ⅳ.V474.1

中国版本图书馆 CIP 数据核字(2020)第 124725 号

**国际空间科学小卫星发展路线图**

GUOJI KONGJIAN KEXUE XIAO WEIXING FAZHAN LUXIANTU

| | |
|---|---|
| **出版** | 中国科学技术大学出版社 |
| | 安徽省合肥市金寨路 96 号，230026 |
| | http://press. ustc. edu. cn |
| | https://zgkxjsdxcbs. tmall. com |
| **印刷** | 安徽国文彩印有限公司 |
| **发行** | 中国科学技术大学出版社 |
| **经销** | 全国新华书店 |
| **开本** | 710 mm×1000 mm　1/16 |
| **印张** | 13.25 |
| **插页** | 8 |
| **字数** | 323 千 |
| **版次** | 2020 年 11 月第 1 版 |
| **印次** | 2020 年 11 月第 1 次印刷 |
| **定价** | 60.00 元 |

# 中译本序

感谢郭世杰和许永建两位译者的努力,使得刚刚发布不久的《国际空间科学小卫星发展路线图》(以下简称《路线图》)能以中文的形式和大家见面。尽管对空间科学和技术领域工作的大部分人而言,阅读英文原文并没有什么障碍,但是,仍然不如有一本自己母语的《路线图》在手随时可以参阅来得方便。

本书是国际空间研究委员会(COSPAR)为了应对小卫星、微卫星和立方星的快速发展以及空间科学研究的需求,提出的一项战略研究项目。项目由国际上相关领域的知名科学家组成的研究组领导和完成。战略研究主要是通过组内的电话/视频会议、集中会议以及几次扩大的研讨会等方式进行,其中两位共同主席做了大量的工作。

其实定义为"小卫星"并不是因为它小。在航天技术发展初期,各国发射的卫星都不大,但是它们不能被称为现代意义上的小卫星。我们这里定义的小卫星是在 20 世纪 80 年代中期出现的,由英国萨瑞大学的马丁·斯威廷(Martin Sweeting)教授发起,通过使用廉价的商用器件,成功地实现了低成本、多用途和搭载发射的小卫星系列,改变了卫星研制的商业生态,极大地推动了遥感、通信和科学实验领域的卫星应用。之后,由于电子技术集成化的发展,这种100 kg 以下的微小卫星的质量、体积进一步缩小,直至形成了质量在 1 kg 左右、体积为 10 厘米见方(10 cm×10 cm×10 cm)的标准化的立方星体系,其研制成本也逐渐降到 100 万元人民币以下。

　　然而这样的微小卫星甚至立方星,除了在大学里发挥了不可替代的教育实习作用以外,在其他方面的应用还是十分有限的,特别是在空间研究方面。而对空间研究来说,由于都是政府项目,经费非常有限,利用廉价的微小卫星甚至立方星就成为了非常吸引人的领域。因此,COSPAR 开展这项战略研究就适逢其时了。

　　尽管这项研究的目的是针对小卫星的,但其覆盖和涉及的相关领域并不少于其他大卫星。因此,尽管研究团队尽了最大努力,仍然不可避免会有一些遗漏。当然,如果读者发现自己的想法并没有被这个路线图包括进去,这自然是一件非常好的事情,甚至可能是独一无二的创新方向。因此,不要被这个路线图描绘的方向所限制。它也仅仅是为大家描述了一个一般性的框架,并为相关管理机构提供了一个计划和安排支持经费的参考基础。相信在这个路线图的牵引下,会有越来越多的小卫星、立方星用于科学研究,推动空间科学在新的维度上快速发展。

吴　季

2020 年 5 月 12 日

# 译者的话

国际空间研究委员会（COSPAR）隶属于国际科学理事会（ISC），作为空间科学领域最大的国际组织，着眼于世界空间科学发展的关键问题，并针对这些关键问题发布科学路线图，为各个领域的学者和决策者提供科学参考。事实证明，这些路线图报告获得了空间科学界和航天机构的认可，COSPAR 在促进空间研究领域的国际合作上发挥了重要作用。自 2012 年起，COSPAR 陆续发布了多份空间科学领域发展路线图，内容覆盖空间探测、空间天文、空间天气、地球系统观测和小卫星发展等。

2019 年 8 月，COSPAR 在《空间研究进展》（*Advances in Space Research*）上发布了《国际空间科学小卫星发展路线图》（以下简称《路线图》）。

《路线图》的编写团队回顾了空间科学小卫星的昨天，盘点了其发展现状，并展望了未来发展，对于参与空间科学小卫星研制的高校、科研部门、航天工业界以及决策者等都有很好的启发意义。具体而言，本书的第 1 章回顾了小卫星的发展历史，定义了小卫星质量的上限为几百千克；第 2 章对小卫星探测在地球和日地空间研究、太阳系探测、合成孔径光学望远镜以及星际探测等领域的发展进行了展望；第 3 章列举了小卫星发展的障碍，如经费支持，配套政策，与工业界的协调、创新和合作障碍等；第 4 章提出了全球小卫星发展的五项建议。

2017 年,《路线图》的起草工作由时任美国密歇根大学教授的托马斯·泽布臣(Thomas Zurbuchen)和国际空间科学研究所(ISSI)主任、伯尔尼大学教授鲁道夫·施泰格(Rudolf Steiger)作为起草委员会共同主席来一同负责。泽布臣担任美国国家航空航天局(NASA)负责空间科学的副局长后,其工作由美国达特茅斯学院教授罗宾·米兰(Robyn Millan)接手。

《路线图》的研究组由活跃在空间科学小卫星领域的知名科学家、工程师和学者组成,具体包括:美国达特茅斯学院的米兰,国际空间科学研究所、伯尔尼大学的施泰格,以色列特拉维夫大学的迈尔·阿里埃勒(Meir Ariel),俄罗斯科学院空间研究所的谢尔盖·巴塔列夫(Sergey Bartalev),欧洲空间局(ESA)的莫里斯·博吉厄德(Maurice Borgeaud),NASA 喷气推进实验室(JPL)的斯特凡诺·坎帕尼奥拉(Stefano Campagnola)和朱莉·罗杰斯(Julie C. Castillo-Rogez),丹麦技术大学的雷内·弗莱隆(René Fléron),瑞士空间中心的沃尔克·伽斯(Volker Gass),意大利里雅斯特大学的安娜·格雷戈里奥(Anna Gregorio),美国蒙大拿州立大学的戴维·克伦帕尔(David M. Klumpar),美国防务分析研究所(IDA)科技政策研究所的巴维亚·拉尔(Bhavya Lal),英国斯特拉斯克莱德大学的马尔科姆·麦克唐纳(Malcolm Macdonald),韩国天文和空间科学研究所的朴钟旭(Jong Uk Park),印度 PES 大学的桑巴瓦·拉奥(V. Sambasiva Rao),德国胡布兰大学的克劳斯·席林(Klaus Schilling),美国斯坦福大学的洛克希德·马丁先进技术中心的格雷姆·斯蒂芬斯(Graeme Stephens),NASA 喷气推进实验室的阿伦·泰特(Alan M. Title)和中国科学院(CAS)国家空间科学中心的吴季。

COSPAR 总部秘书处对本书的出版给予了最大限度的支持。在《路线图》中文版出版之际,译者团队对英文版编写团队和 COSPAR 秘书处表示衷心的感谢。COSPAR 前副主席、中国科学院国家空间科学中心的吴季研究员作为英文版编写团队成员和领域专家为本书作序推荐。中国科学院国家空间科学中心的李靖研究员对本书进行了细致审校。哈尔滨工业大学卫星技术研究所授权使用搭载嫦娥四号中继星发射的龙江一号和龙江二号环月微卫星图片作为封面素材。在此一并表示感谢。

译者团队曾先后参加了以小卫星为主题的第三届 COSPAR 研讨会

（2017 年，韩国济州）、第四届 COSPAR 研讨会（2019 年，以色列荷兹利亚）以及第 42 届 COSPAR 大会（2018 年，美国帕萨迪纳），见证了《路线图》起草的重要节点。希望《路线图》中文版有助于对空间科学小卫星感兴趣的中国读者了解国际空间科学小卫星发展的前沿领域，从而助力国内小卫星事业的发展。

译 者

2020 年 5 月于北京

# 前　言

　　《国际空间科学小卫星发展路线图》由国际空间研究委员会（COSPAR）负责制定，旨在通过利用小卫星的创新和国际合作促进空间科学前沿研究。全球小卫星领域的快速发展使我们有机会利用小卫星的发展形势来实现科学进步。具体而言，低成本发射机会正变得越来越多，商业硬件的普及也降低了空间科学小卫星任务的成本。这反过来会提高小卫星的发射频率，鼓励科学家提出更多创新性的小卫星任务建议，并最终促成科学突破的实现。此外，新计算机技术和计算方法正在改变数据的获取、管理和处理的方式，而小卫星所带来的大量数据集客观上要求科学数据分析采用新的范式。本书举例说明了由小卫星发展变革所带来的长期科学愿景。考虑到制定路线图的目的，文中的"小卫星"并不十分严格地被定义为质量不超过几百千克的航天器，但是研制和发射这些航天器的过程比这一质量限制指标更加重要。本书旨在鼓励空间科学界充分利用小卫星工业界的新发展形势来增加卫星任务发射频率，并改变空间科学小卫星的研制和管理方式。最后，本书分别向空间科学界、航天工业界、各国航天局、航天政策制定者和国际空间研究委员会提出了一些建议。

　　2017 年初，在国际空间研究委员会的支持下，由领军科学家和工程师组成的小卫星国际研究小组正式成立，开展一项为期两年的研究来制定《国际空间科学小卫星发展路线图》(4S)①。本书定义"小卫星"的质量上限为几百千克，但是作为(判定)指标，卫星质量的重要性低于研制和发射这些卫星的过程。由于立方星(CubeSats)在小卫星发展变革中起到了关键作用，因此本书也对立方星和利用立方星技术的小卫星进行了重点讨论。立方星是小卫星的一种，以10厘米见方为一个单位(一个单位称为1U，两个单位称为2U，以此类推)。此外，本书的研究受到两份研究报告的启发：一是泽布臣等人于2016年发表在《今日空间研究》上的论文，二是美国科学工程和医学院(NASEM)的一份战略报告②。

空间科学小卫星路线图委员会的任务是回答如下6个问题：

(1) 小卫星发展的现状和利用情况如何？尤其是用于空间科学研究的立方星，其技术能力和迄今为止取得的主要成就有哪些？

(2) 不管小卫星是作为独立的发射任务，还是作为搭载载荷，或者作为小卫星星座和集群，其科学潜力如何？

(3) 在航天器研制(包括硬件和软件)以及地面系统的研发标准化方法方面，参与的航天机构和工业界等部门的作用是什么？

(4) 立方星及利用立方星技术的小卫星数量和种类的增长得到了哪些政策支持？与之相关的通信和频率分配、轨道碎片和运载火箭的政策支持有哪些？

(5) 负责小卫星研制和小卫星任务运行的团队之间的国际合作有什么成功模式？未来如何共享数据并进行数据储存？

(6) 小卫星领域发展迅速，参与小卫星项目的高校和相关国际组织应该如何相互学习、分享经验并推动国际合作？

本书致力于回答如上问题，以有益于对促进全球小卫星任务发展抱有兴趣的各国航天机构及其背后的政府、非营利组织和其他私有组织。此外，我们希望空间科学界积极利用小卫星工业界的新发展形势，积极与工业界合作，积极拓展国际合作，以增加小卫星任务的发射频率，并改变空间科学小卫星的研制和管理方式，最终实现空间科学的巨大发展。《路线图》的研究小组成员的学科背景涵盖了天基观测的各个领域，其中包括地球科学、太阳和空间物理、行星科

---

① 缩略词的具体含义见缩略语部分。

② 美国科学工程和医学院的战略报告详见 https://www.nap.edu/catalog/23503/achieving-science-with-cubesats-thinking-inside-the-box。

学和天文学等。研究小组成员中既有科学家、工程师,也有政策专家,他们来自高校、公立研究机构和工业界。

　　本书的指引作用在一定程度上类似于很多山地徒步景区或者滑雪度假区的路径指示图。可以进行如下类比:前景中描绘的村庄是"我们的(小卫星)社区";而远处标识为禁地的高山、险峰则代表需要很长时间才能实现的愿景和目标;图上的路径系统将引领我们闯过危险的山区。以此为蓝本,本书主要分成了如下3个部分:第1章"我们的小卫星社区",介绍了小卫星和立方星的历史和目前的发展情况,其中包括技术现状和近期的科学潜力。第2章"小卫星未来展望",描述了在更遥远的未来(未来十年及更远)能够实现的科学卫星任务。虽然最终提出未来空间科学卫星任务概念和未来发展重点的是空间科学界,但是本章中概述的展望内容将有助于我们专注于对小卫星的讨论。第3章"小卫星未来发展的障碍和对策",描述了一些制度上的障碍和克服它们的方法,并着重讨论了航天机构和工业界的职责、国际合作政策和国际合作模式等。

　　我们最终的追求是:由科学家组成的国际科学团队努力实现具有创新意义和深远影响的科学目标。本书提供了实现这一目标的一些可能途径,其研究的一些主要发现将在各个章节中列出,并基于这些主要研究发现,提出了如下5项建议:

### 1. 对科学界

　　整个科学界应该认同小卫星的价值,并寻找机会利用小卫星工业界的新发展形势。空间科学各个领域都可能从更小尺寸、更低成本和更快迭代周期的小卫星发展中获益,尤其是小国家的空间科学界会受益于对小卫星的投资。

### 2. 对航天工业界

　　卫星开发企业应该主动寻求与科学家、高校和大型政府机构的合作机会。具体而言,合作机会可能涵盖数据共享协议、出售商业航天器上可以搭载科学载荷的空间等。目前,开放数据对实现科学目标而言有非常高的价值。商业公司应该对科学界需要使用的完整和开放的数据协议持开放态度,这种合作关系也有助于相关人力资源的发展。

### 3. 对航天机构

　　大型航天机构应该根据小卫星项目的体量制定合适的步骤和程序。各国航天机构应该探索新方式为科学、应用和技术验证这三类小卫星提供发展机遇和充足的发射窗口。此外,各国航天机构应该充分利用商业数据和商业航天设施来进行科学研究,并遵循数据开放政策。最后,各国航天机构应该携手制定小卫星发展长期路线图,以明确未来优先发展的国际合作小卫星任务。

### 4. 对政策制定者

为了确保空间科学小卫星任务的成功,科学界需要政策制定者的如下支持:① 确保频谱的充分可用,采取轨道碎片减缓和整治方案,以及提供经济上可负担的发射机会和其他基础设施服务;② 确保制定的出口管制指南易于理解和解释,并做好国家安全和科学利益之间的平衡;③ 针对小卫星频谱获取、轨道机动性、可跟踪性和寿终报废处置等方面的国家和国际法规开展教育和指导。

### 5. 对国际空间研究委员会

国际团队可以共同为类似于欧盟"50 小卫星组网"(QB50)这样的模块化国际小卫星星座任务确立科学目标和规则,而 COSPAR 在这个过程中应起到促进作用。通过 1957~1958 年"国际地球物理年"(IGY)这样的活动,参与者可以就小卫星发展的基本规则达成一致,而各国航天机构或相关机构代表应该从一开始就参与进来。资金来源于各成员国,甚至可以来自私人实体或基金会。COSPAR 并不提供资助,而是扮演忠实的居中协调角色,并规定这些国际团队在提出任务建议时必须遵守的准则。

通过国际合作来研制小卫星星座(small satellite constellations)对所有任务参与方来说都是十分有价值的,并且总体价值高于各国"单打独斗"产生的价值之和。这种大规模的行动计划将有利于实现愿景目标,并最终在技术和科学上实现突破。小卫星任务能够催生新的国际合作模式,使更多国家可以在世界范围内参与雄心勃勃的航天项目。COSPAR 的角色是促进国际合作,以开放的姿态为搭建"社区公民科学"创造典范。我们最后提出的这条建议是促进真正宏大想法付诸实践的一种途径,这些宏大想法包括本书所谈到的 4 个未来愿景以及其他目前还未想象到的宏大想法。

# 目 录

<h1>第 1 章</h1>

<h1>我们的小卫星社区</h1>

本章概述小卫星的历史、现状和小卫星近期的科学潜力。

## 1.1　小卫星和立方星的历史和现状

小卫星工业的发展日新月异。本节扼要介绍小卫星发展的历史和小卫星工业的发展概况。对小卫星历史更加全面的回顾可以参考斯威廷于 2018 年发表的综述(Sweeting,2018)。

### 1.1.1　传统的空间科学小卫星

质量大约在 100 kg 以上的小卫星已经完全证明了其在空间科学卫星任务中的用途,并成为近几十年来空间科学知识增长的主要贡献者,尤其体现在太阳物理学[①]、天体物理学和地球科学这些子学科中。

在美国,大部分空间科学小卫星任务均由美国国家航空航天局(NASA)的"探索者计划"(Explorers Program)支持,该项目为天体物理和太阳物理方面的科学探索卫星计划提供发射机会。自 1958 年发射"探索者 1 号"以来,该项目已经支持

---

　[①]　太阳物理学包括对行星际介质空间环境的研究,也包括对太阳、日球层、地球空间以及太阳系与星际空间之间相互作用的研究。

了 70 多个空间科学卫星任务(包括美国本土的任务和国际合作卫星任务),一共发射了 90 多颗卫星。"探索者计划"已经在如下领域交出了令人印象深刻的成绩单:地球磁层和重力场的形状,太阳风,落到地球上的微流星体的性质,来自太阳系及其外部的紫外线、宇宙射线和 X 射线的许多知识,电离层物理,太阳等离子体,高能粒子和大气物理等。这些空间科学任务还研究了大气密度、射电天文学、大地测量学和伽马射线天文学。一些探索者航天器甚至已经飞行到其他行星,有些已经对太阳开展了监测。[①] 1958～1962 年发射的早期探索者卫星质量不到 50 kg。随着探测能力的迅速提高,卫星质量也水涨船高。"探索者计划"的第 66 次发射任务是 1989 年发射的"宇宙背景探测器"(Cosmic Background Explorer,COBE),其卫星干重已达到 1408 kg。1988 年,NASA 启动了"现代探索者计划",将小型空间科学航天器的质量控制在 60～350 kg,此举是为了应对卫星质量的迅速增加以及随之而来的成本上升和发射频率下降等问题。有大约 17 颗卫星的质量在这个范围内,其中包括"大学探索者计划"(University Explorer line,UNEX)卫星、"小型探索者计划"(Small Explorer line,SMEX)卫星和一次发射 5 颗的"中型探索者计划"(Medium Class Explorer,MIDEX)卫星。其中"小型探索者计划"的 10 颗卫星的发射质量见表 1.1。

表 1.1  NASA"现代探索者计划"中小卫星的发射质量

| 任务名称 | 质量(kg) |
| --- | --- |
| SAMPEX | 158 |
| FAST | 187 |
| TRACE | 250 |
| SWAS | 288 |
| RHESSI | 230 |
| GALEX | 280 |
| AIM | 197 |
| IBEX | 80 |
| NuSTAR | 350 |
| IRIS | 200 |

注:数据来自 https://nssdc.gsfc.nasa.gov/multi/explorer.html 网站,AIM 卫星和 NuSTAR 卫星的质量数据来自其任务网站。

---

① 探索者计划历史:https://explorers.gsfc.nasa.gov/history.html。检索日期 2018 年 5 月 27 日。

就科学回报来说,"探索者计划"是非常成功的。但是,该计划的发射频率并没有得到充分的提高。自 1958 年至 1980 年以来,探索者计划一共发射了 62 次(2.82 次/年);而在 1980 年(航天飞机元年)至 2018 年期间,探索者计划只发射了 33 次(0.87 次/年),发射频率降低了三分之二以上。最近两次太阳物理任务的发射间隔是 11.5 年:2002 年 2 月,"拉马第高能太阳光谱成像仪"(RHESSI)发射;2013 年 6 月,"界面区成像光谱仪"(IRIS)发射。IRIS 是最近的一次小型探索者发射任务,距今也已经过去 7 年多了。加强监督管理可能导致了卫星研制时间的增加。另外,高昂的发射成本也可能是发射频率下降的一个重要因素,NASA 于 2018 年发射了质量为 362 kg 的中型探索者任务"凌日系外行星勘测卫星"(Transiting Exoplanet Survey Satellite,TESS),其发射成本可以作为参考:该卫星耗资 2 亿美元(Wall,2018),但不包括 8700 万美元的发射成本[①],其中发射成本的占比超过任务总成本的 30%。

最近 NASA 确立了"地球探险"(Earth Ventures)系列卫星计划,为地球科学的小卫星任务提供机会。其中,"飓风全球导航卫星系统"(Cyclone Global Navigation Satellite System,CYGNSS)是"地球探险计划"的第一个卫星任务,主要用于海洋风测量,该任务验证了小卫星星座在地球科学领域的应用。此卫星任务由 8 颗同时运行的小卫星组成,每颗卫星的质量约 28 kg。CYGNSS 任务于 2012 年立项研制[②],并于 2016 年发射[③]。

欧洲的小卫星研发通常能获得不同层级计划的支持,包括国家计划、欧盟第七框架计划(7th Framework Program,FP7)、"地平线 2020"(Horizon 2020)战略计划以及 ESA 的计划。戴尔(Dale)和惠特科姆(Whitcomb)于 1994 年概括介绍了 ESA 小卫星的历史(Dale,Whitcomb,1994):在欧洲,小卫星任务最初被归为空间科学计划(参见 1985 年制定的《"地平线 2000"战略计划》)。当时,ESA 正在研制星簇任务(Cluster mission),考虑采用与 NASA 的"磁层粒子主动示踪探测器"(Active Magnetic Particle Track Explorer,AMPTE)类似的采购规则。但最终结论是:"如果对星簇计划采用'小卫星'的管理方法,其变数太大,因此应该按照更'经典'的流程进行研制。"1990 年,ESA 发出了小卫星任务建议征集,并对收到的 52 项任务建议书进行了评估。其中遴选出两个任务进行了进一步的研究:"太阳

---

① https://www.nasa.gov/press/2014/december/nasa-awards-launch-services-contract-for-transiting-exoplanet-survey-satellite,检索日期 2018 年 6 月 26 日。

② https://www.nasa.gov/home/hqnews/2012/jun/HQ_12203_Earth_Venture_Space_System_CYGNSS.html。

③ https://www.nasa.gov/image-feature/cygnss-satellites-launchedaboard-pegasus-xl-rocket。

扁率、辐照度周期和直径变化任务"(SOLID)和"宇宙紫外线背景勘测任务"(CUBE)。1992年11月,ESA发布了针对小卫星任务建议书的具体要求,大约13个任务建议书进入了任务评估阶段,但是最终没有一项获得推荐进入进一步研究环节。报告的结论是:"虽然ESA科学项目部对引入小卫星项目的实用性和发展潜力没有形成固定政策,但已普遍认识到降低卫星任务总成本的必要性,这样可以增加卫星发射机会和小型卫星计划的数量。"该报告还指出,较小国家可能没有设计、研制和发射小卫星任务所需的基础设施。因此,ESA可以提供潜在的发射机会,或作为国家级卫星项目的"中间人"。事实上,比利时的"星上自主项目"(Project for On-Board Autonomy,PROBA)系列微卫星就是由ESA的小卫星项目资助的。其中PROBA-1于2001年发射,卫星质量为94 kg,它是飞行时间最长的地球观测任务①。PROBA-2和PROBA-5卫星也在轨多年,分别收集了关于太阳活动和植被/土地利用情况的数据。ESA的首个月球任务"智慧一号"(SMART-1)的质量刚刚超过350 kg,它搭载地球同步转移轨道卫星发射(2003年9月),并测试了太阳能电推进技术②。

2012年,ESA宣布将资助一系列常规的小卫星任务,也就是"S级"任务,其部分目的是为较小成员国提供主导卫星任务研制的机会③。在第一次小卫星任务建议书征集通知发布后,ESA收到了大约70份任务意向书④,这也显示了ESA成员国对小卫星任务的巨大兴趣。2012年遴选出了"系外行星特征卫星"(CHaracterising ExOPlanets Satellite,CHEOPS),该任务计划于2019年发射。第二个"S级"任务为"太阳风-磁层相互作用全景成像卫星"(Solar wind Magnetosphere Ionosphere Link Explorer,SMILE),也称为"微笑计划"。该任务是ESA与中国科学院的联合任务,旨在研究太阳风、磁层及电离层之间的相互作用。尽管属于"S级"任务,但"微笑计划"的卫星质量并不是很小;卫星搭载了4种主要载荷,卫星干重为652 kg,如果加上推进剂,总质量为1960 kg(Raab et al.,2016)。与NASA的"探索者计划"类似,ESA对100 kg左右的卫星任务提供的机会比较有限,其任务研制周期接近10年。

ESA只是偶尔参与空间科学微卫星项目,而ESA成员国往往是以国家为基

---

① https://www.esa.int/Our_Activities/Observing_the_Earth/Proba-1。

② http://sci.esa.int/smart-1/38890-smart-1-mission-to-the-moon-statusfirst-results-and-goals。

③ https://www.bbc.com/news/science-environment-17335339。

④ http://sci.esa.int/cosmic-vision/50265-received-letters-of-intent/,检索日期2018年7月8日。

础,为小/微卫星提供了更多发射机会,并开放了小卫星领域,使加入任务的国际伙伴能够从经济上可承受。自 1988 年至 2016 年,ESA 只发射了 6 颗质量低于 200 kg 的卫星,而 ESA 成员国在这期间共发射了 131 颗小卫星(图 1.1,参见彩图)①。例如,在 20 世纪 90 年代早期,瑞典和德国联合发射了研究极光的 Freja 卫星任务②,该卫星质量为 214 kg。其他小卫星领域的参与者还包括丹麦。丹麦成功酝酿、设计、研制了质量为 61 kg 的 Ørsted 卫星任务,并成功实现了对卫星的运行控制,该卫星任务于 1999 年发射。Ørsted 卫星任务旨在对地磁进行探测和测绘,为地球内部发电机(dynamo)的研究提供了信息(Hulot et al.,2002),提高了我们对电离层和磁层电流系统的理解(Papitashvili et al.,2002),为国际地磁参考场(IGRF)模型提供了 5 年的数据来源。法国国家空间研究中心(CNES)已经发射了一定数量的小卫星,并研制了可同时用于地球科学卫星任务和军事卫星任务的 100 kg 的多用途米里亚德(Myriade)卫星平台。法国发射的首颗小卫星是 2004 年发射的"迪米特"(Demeter)卫星,最近的一次任务是于 2019 年发射的"塔拉尼斯"(Taranis)卫星③。最近,意大利航天局(ASI)启动了一项名为"PLATiNO"的小卫星计划,旨在通过开发多用途小卫星平台,建立国家空间科学任务和其他任务的能力④。

俄罗斯有很长的小卫星发射历史,这始于 1957 年发射的第一颗人造地球卫星"斯普特尼克 1 号"(Sputnik-1)。俄罗斯的高校一直热衷于研制空间科学小卫星。例如,"达吉雅娜-2"(Tatyana-2)卫星就是一颗由国立莫斯科大学主导的国际合作微卫星(约 100 kg),该卫星于 2009 年作为搭载有效载荷发射,用于研究地球大气中的瞬态发光事件。

在其他一些国家,小卫星主要用于工业或业务运行,而不是用于科学。例如,在日本,负责空间科学的日本宇宙航空研究开发机构(JAXA)一直在研制类似于"光学导航临近天体近距离飞越任务"(Proximate Object Close flyby with Optical Navigation,PROCYON)的技术验证的微卫星,但 JAXA 的主要工作是研制大型卫星。小卫星主要被视为用于推动工业发展和改善生活的手段(如利用小卫星进行对地观测应用,包括预报海啸等)。举例来说,日本北海道大学和东北大学最近

---

① 这些卫星任务中的部分任务可能获得过 ESA 的项目资助。

② https://nssdc.gsfc.nasa.gov/nmc/spacecraftDisplay.do?id=1992-064A。

③ https://myriade.cnes.fr/en/home-49。

④ 多用途小卫星平台由 SITAE 和意大利航天局共同规划。http://www.satnews.com/story.php?number=275392102。

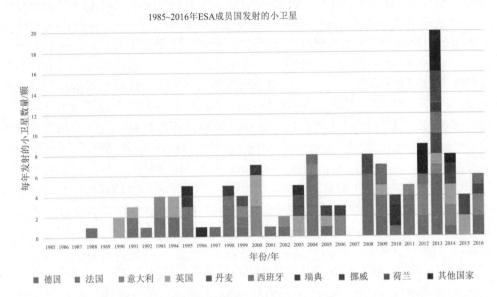

**图 1.1　1985～2016 年 ESA 成员国发射的小卫星(小于 200 kg);相比之下,ESA 同期发射了 6 颗小卫星[数据收集由拉尔等人(Lal et al.,2017)完成,数据由拉尔提供]**

启动了一个项目,计划到 2020 年发射 50 颗用于自然灾害监测的微卫星①。该项目也有区域国家的参与。重量为 50 kg 的"迪瓦塔-1 号"(Diwata-1)是菲律宾自主研制的第一颗人造卫星,于 2016 年 4 月从国际空间站释放后进入轨道②。在过去的 5 年里,大部分小卫星都是由美国发射的。但是,越来越多的国家正在研制小卫星(图 1.2,参见彩图)。尤其是中国发射的小卫星数量很多,但大多用于国防或工业。对当前国际小卫星项目的全面总结可参阅拉尔等人的文章中的附录 E(Lal et al.,2017)。

从世界范围来看,用于空间科学的小卫星数量只占到小卫星发射总量的很小一部分;大多数小卫星任务用于遥感或技术研发(图 1.3,参见彩图)。质量低于 200 kg 的小卫星有巨大的科学潜力,下文将进一步讨论。

---

① http://www.satnews.com/story.php? number=900912903,检索日期 2019 年 3 月 4 日。

② https://www.rappler.com/science-nature/earth-space/130956-diwatamicrosatellite-deployment-space,检索日期 2019 年 5 月 27 日。

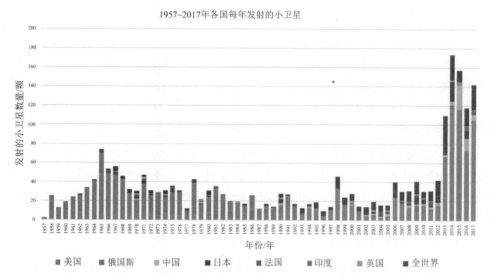

图 1.2　1957～2017 年各国每年发射的小卫星(小于 200 kg)数量[本图转载自 Lal 等人 (2017)的报告"全球小卫星趋势"中的图 E-2,并已经过防务分析研究所同意]

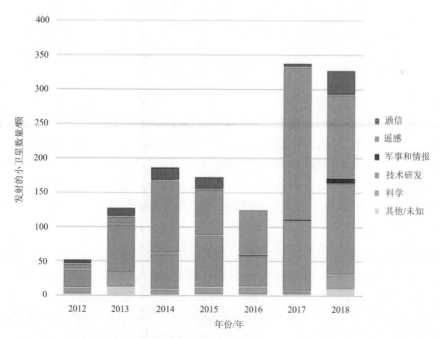

图 1.3　各种用途的小卫星数量[图片来源:Bryce Space and Technology (https:// brycetech.com/reports.html)]

## 1.1.2 立方星

在过去的 15 年里,一种被称为立方星的新类型卫星数量呈爆炸式增长,这种卫星质量在 1~12 kg。其最初的作用是对大学生进行技术培训,如 2009 年发射的"瑞士立方星"(SwissCube)(Noca et al.,2009)。截至 2018 年底,已发射的立方星总数大约为 1030 颗①。

立方星属于纳卫星(nanosatellite),通常是以标准化容器的形式发射并部署到太空中,通常作为大航天器的二级甚至三级有效载荷搭载发射。其得名源于最初的外形:10 厘米见方(10 cm×10 cm×10 cm)的立方体(即 1U)。立方星容器的标准化有利于研制出不同 U 倍数的立方星,其大小可以是:0.5U、1U、1.5U、3U 和 6U,总发射质量可以高达 12 kg。

《路线图》回顾了立方星爆炸式增长的发展历程,并聚焦其科学应用价值。图 1.4(参见彩图)显示了 2000~2018 年全球立方星以及更早出现的纳卫星的年发射频率。自 2014 年起,由 Spire 和 Planet 卫星项目发射的 450 多颗立方星没有计算在内,其发射数量远远超过了其他卫星项目。大多数立方星用于教育、技术验证或商业。截至 2018 年 12 月,累计发射的 1000 多颗立方星中被认定为科学驱动的立方星有 107 颗,其中有些立方星传回了高质量的数据,揭示了空间环境某些方面的特征,这些特征此前尚未被大卫星很好地研究过。

图 1.5(参见彩图)展示了空间科学立方星的年度发射量。其中大约一半空间科学立方星是在 2017 年初以后发射的,其中包括了 QB50 星座计划里的 36 颗卫星。对于该星座计划,下面将详细讨论。107 颗空间科学立方星中有 49 颗已宣布获得成功。也就是说,这些任务的主要目标已经实现,或者卫星正在努力实现主要科学任务目标。然而,最近发射的一些立方星仍然处于在轨测试或早期运行阶段,它们的科学产出仍有变数。因此,立方星任务预期成功数量会随着时间的推移而逐渐增加。

作为欧盟组织的 QB50 星座计划的一部分,36 颗立方星于 2017 年发射升空,并对高层大气进行探测②,很多国家/地区都向该星座计划提供了立方星,其中包括澳大利亚、美国、加拿大、中国、韩国、以色列、南非、土耳其和乌克兰。欧盟的QB50 项目管理部门为各国的立方星任务小组提供了广泛的技术和管理支持,包括

---

① Erik Kulu, Nanosatellite & CubeSat Database, https://www.nanosats.eu/。
② https://www.qb50.eu/。

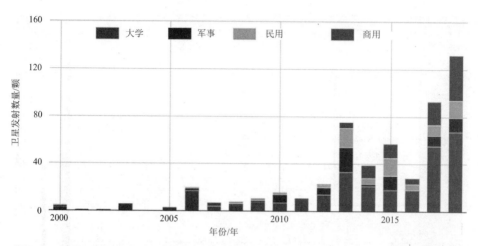

**图 1.4**　**2000～2018 年每年立方星的发射数量,根据负责设计/建造/运行卫星的机构类型划分(大学立方星、军事立方星、民用立方星、商业立方星)。本图不包含自2014 年开始由 Planet 和 Spire 项目研制的 450 多颗商业星座立方星[图表由 M. Swartwout 制作,使用数据截止到 2018 年底(数据来源:https://sites.google.com/a/slu.edu/swartwout/home/cubesat-database)]**

专业设计评审、科学有效载荷研制和完整的发射活动服务。一些立方星任务团队还免费得到了完整的姿态测定与控制系统(Attitude Determination and Control System,ADCS)。这些团队需要另外投入 60 万～70 万欧元。QB50 项目可以被视为国际联合星座任务的探路者(pathfinder)。参与国要为联合星座任务提供一个完整的航天器,而不仅仅是提供单一的有效载荷或分系统,3.5.1 节将对其进一步讨论。

大多数美国空间科学立方星任务都由美国国家科学基金会(NSF)支持。此前,NASA 的立方星任务都聚焦空间技术的发展,直到最近才发生转变。在 NSF 的立方星任务取得成功之后,NASA 增加了对空间科学立方星的资助。在撰写本书时,NASA 已发射或正在研制的空间科学立方星有 20 多颗。

欧洲对空间科学立方星任务的兴趣也持续增长。自 2005 年起,欧洲小卫星项目主要用于教育实践,卫星任务的雏形是高校研制的 1U‐3U 立方星。2010 年,欧洲开始转向研制更大的 3U‐12U 立方星,研制主体变成了工业界和欧洲各国航天机构,其目的也演变成为技术验证。2009 年,德国发起了面向高校的教育类立方星项目,该项目已发射 6 颗立方星,并有很多处于准备阶段的立方星任务。目前,爱尔兰正在研制本国第一颗卫星——一颗测量宇宙伽马射线爆发的 3U 科学

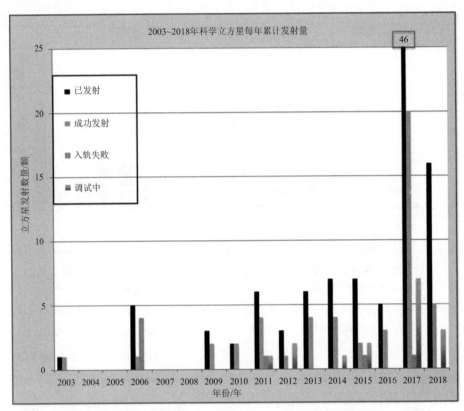

图 1.5　科学立方星：2003～2018 年累计发射了 107 颗；仅 2017 年就发射了 46 颗。107
　　　　颗立方星中的 7 颗因为发射失败而未能入轨。图中标注 4 个类别：已发射立方
　　　　星、成功立方星、入轨失败、立方星调试中。另外 2 个类别未体现在图中，分别
　　　　为到达预定位置时已失灵（此类共计 21 颗）和立方星状态未知（此类共计 14
　　　　颗）（数据来源：Swartwout，https://sites. google. com/a/slu. edu/swartwout/
　　　　home/cubesat-database）

立方星①。自 2013 年以来，ESA 为其"一般支持技术计划"（General Support
Technology Programme，GSTP）安排了超过 1000 万欧元的预算，用来支持 7 个在
轨验证（In-Orbit Demonstration，IOD）立方星任务。其中包括几个用来验证二氧
化氮污染物监测、天气预报和空间天气等技术的近地轨道星座任务。一项"Ka 波
段干涉测量集群"（Ka-band Interferometry Swarm，KRIS）小卫星任务将验证测
量洋流和海面高度的能力。这些验证卫星还包括一颗测绘月球冰图和研究流星影
响的探月立方星，以及一颗独立的深空立方星——微型小行星远程地球物理观测

---

① https：//www. rte. ie/news/business/2017/0523/877210-satellite/。

器(Miniaturised Asteroid Remote Geophysical Observer, M-ARGO)(Walker, 2018)。欧盟委员会在其《程序工作项目 2018—2020》中明确提到了小卫星任务：只要有利于空间科学和使探测任务取得显著的科学产出，就应该鼓励使用立方星和其他小航天器平台，或者使用商用货架产品(COTS)等创新方法。毫无疑问，立方星的崛起加速了小卫星在空间科学领域的应用。

> **研究发现 1.1** 从立方星到 300 kg 左右的小卫星，各种大小的小卫星使空间科学各个领域重要的科学进展成为可能。

> **研究发现 1.2** 小卫星，尤其是立方星，成为更多国家进军太空的窗口；新的小型空间项目为这些国家参与更大的国际项目提供了机会。

> **研究发现 1.3** 立方星的出现显著增加了发射频率。然而，在过去的几十年里，传统的、相对较大的小卫星任务的发射频率下降了，而卫星研制时间和成本并没有相应下降。

## 1.1.3 发射机会、商业化和其他发展

有限的发射机会是早期限制小卫星发展的因素之一。19 世纪 60 年代出现了第一次搭载发射的热潮，但随后数年这种搭载发射机会又变成了稀有资源。2007 年前后由于立方星的标准化和多种发射平台的出现，搭载发射的数量大幅增加。标准化使得携带立方星的容器可以搭乘十几种不同的运载火箭和航天器，其中就包括国际空间站。这种标准化加上相对较低的立方星研制和发射成本，使立方星超出了教育和培训的范围，最终形成了在应用领域爆炸式的增长。最近，商业界也开始利用立方星组成全球星座，每个星座的立方星数以百计。

商业部门对小卫星兴趣的增长有望带来新的发射机会，并降低发射成本。拉尔等人对市场驱动因素和发射机会进行了综合分析，可参阅拉尔等人于 2017 年发表的文章(Lal et al.,2017)的第 3 章和第 4 章。现在，进入近地轨道(LEO)的发射机会相对便宜，并且发射密度也相对更大。这些机会包括从国际空间站的释放以及利用各个国家和地区火箭的搭载发射机会，如印度的极轨卫星运载火箭(PSLV)、俄罗斯的第聂伯(Dnepr)运载火箭和宇宙(Cosmos)运载火箭、中国的长征系列火箭、ESA 的织女星(Vega)小型运载火箭和美国的猎鹰(Falcon)运载火箭

等。有时，一发火箭也搭载许多小卫星一起发射，例如印度的极地轨道运载火箭于2017年2月将104颗小卫星送入了太空。此外，还有一些新航天公司为小卫星设计和生产火箭，如火箭实验室（Rocket Labs）和维珍轨道（Virgin Orbit）公司。一些经纪机构可提供航天器打包服务，通过联系具有送入预定轨道发射运力的航天机构来寻找发射机会。更详细的发射相关政策将在3.2.3节中展开讨论。

小卫星的商业化和新航天公司的入局同样也促进了完整的商业货架子系统的供应，而这些商业货架产品的供应有可能大大减少空间科学卫星的成本和研制时间。商业部件已用于科学立方星，并且大规模生产的部件的可靠性也已经得到了证明。例如，空间创新解决方案公司（Innovative Solutions in Space，ISIS）[①]已经为260个小卫星任务提供了器件和子系统。NASA的小型航天器虚拟研究所（Small Spacecraft Virtual Institute）[②]现在为用户提供了一个零部件搜索工具，通过这个搜索工具，用户可以获取有关商业器件和部件的可靠性信息，这些信息将有助于决定能否在比立方星更大的科学卫星上使用这些器件和部件。

小卫星发展可以为空间科学小卫星创造一种新模式。低发射成本、商业货架产品和购买整星的能力等因素的结合，可以使任务成本降低到一个临界点，并使流水线测试、结构验证和分析的成本更低。此外，小卫星测试可以采用更小型的设施，从而也能降低成本。商业航天部门已经开始研发小卫星制造和测试的新方法。例如，英国一网（OneWeb）卫星公司最近新建了生产线，采用飞机制造技术每天最多可以生产3颗卫星[③]。一网卫星公司已经以低于每颗100万美元的价格预订了900颗质量为150 kg的小卫星（Lannotta，2019）。

> **研究发现1.4** 立方星发射频率的快速增长可以归功为3个原因：标准化的实施增加了搭载发射机会；采用商业货架产品降低了成本；在私人航天领域的应用获得了爆发式增长。

> **研究发现1.5** 越来越多的搭载发射机会和更大的运载火箭带来的成本效益，加上更小的航天器和低成本的商业货架产品，使得大型星座计划（如Planet计划和QB50计划）成为可能，这为空间科学提供了新机遇。

---

① ISIS, Motorenweg 23, 2623 CR, Delft, The Netherlands, https://www.isispace.nl.

② https://www.nasa.gov/smallsat-institute.

③ http://www.oneweb.world/press-releases/2017/oneweb-satellites-breaks-ground-on-the-worlds-first-state-of-the-art-high-volume-satellite-manufacturing-facility.

在过去的 30 年里,科技进步彻底改变了人们的生活、工作和驾驶方式。新技术给我们带来了互联网、智能手机和安全系数更高也更加智能的汽车,然而这些新技术并没有应用在空间任务中。相反,大多数空间科学任务的部件都是十多年前就有的。对造价昂贵的大型卫星任务来说,采用具有高可靠性的有资质部件是有必要的。低成本的卫星任务风险承受能力更强,可以利用还未经过多轮空间验证的最新技术。就目前而言,只有立方星在应用这样的新技术,而传统的、较大的空间科学小卫星(约 100 kg)还未采用最新的技术。

因此,传统的空间科学小卫星任务没有实现成本的降低和研发时间的减少,而商业航天领域却做到了成本和研发时间的"双降"。自 1988 年 NASA 重新启动"探索者计划"以来,小型卫星任务从遴选到发射的平均周期为 5.6 年。这不包括 NASA 发布任务建议书征集到开始遴选的时间(大约 9 个月),也不包括形成任务概念的准备时间(大概需要几年时间),而任务概念必须足够成熟才有可能进入遴选阶段。总而言之,从载荷概念到开始科学分析大约需要 7.5±2 年。

如果空间科学小卫星任务时间周期太长,成本太高,将导致卫星任务建议书征集间隔过久,并且建议书成功率也会比较低。而上述情况又加剧了任务遴选过程中的风险规避,这种风险规避甚至也会出现在由国家航天机构主导的小卫星任务遴选过程中。科学家可能会不愿意进行创新和承担风险,这是因为科学家在整个职业生涯中可能只有一两次机会来领导一项卫星任务。因此,这种风险规避可能会导致平庸或按部就班的科学产出。而商业小卫星的繁荣为小卫星的发展提供了一种新方法,这也是改变空间科学小卫星范式的一个难得的机会。

> **研究发现 1.6**　空间科学界还未能充分利用技术进步和商业航天日益增长的活动,以降低传统小卫星的成本和缩短研发时间。发射机会不足的问题仍然存在,这可能会阻碍航天机构的出资和科学家的创新。

## 1.2　小卫星和立方星的科学潜力

本节聚焦小卫星的短期科学潜力,并重点介绍目前正在论证的一些卫星任务概念。但本节并不是对小卫星和立方星的全面介绍,而是对目前使用小卫星进行广泛科学应用的说明。此外,还将讨论目前利用小卫星进行空间科学研究的局限和挑战。

## 1.2.1　概述

传统小卫星的重要性已得到空间科学界的认可和重视,特别是在天体物理学、太阳物理学和地球科学领域。涉及这些学科领域的最近一份《十年调查》(*Decadal Survey*)[①]建议,继续加强 NASA 的传统小卫星计划(即"探索者计划"和"地球探险计划")。除此之外,类似于立方星的更小卫星计划将有助于实现新的空间科学发展。2016 年,美国国家科学院的报告《用立方星实现空间科学》(*Achieving Science with CubeSats*)全面概述了立方星的科学潜力(NASEM,2016)[②]。该报告的结论是:立方星不会取代大型卫星任务;与此相反,立方星可以用来实现既定科学目标,并通过提供辅助测量等手段助力于更大的卫星任务。报告指出,在太阳和空间物理学领域,立方星可以提供创新测量方法,例如在高风险轨道助力大型设施并具备实现星座任务的潜力。星座任务在地球科学领域也有重要的应用。由于研发时间较短,立方星也有能力填补长期地球观测任务的空白,并具有对新观测需求反应快速的特点。在天体物理领域,立方星的大小限制了观测孔径,从而也限制了单一立方星能够实现的科学目标。但是,该报告突出强调了一些立方星的能力,包括在系外行星研究和恒星变率研究中对单一天体进行长时间观测的能力。立方星星座也有可能为空间干涉测量奠定基础。在行星科学领域,立方星可以提供独一无二的有利位置对高风险区域进行探测,并有可能和更大的"母星"协同合作。此外,立方星还可以作为微重力实验室。报告还指出,立方星已经在上述领域实现了高影响力的科学成果。

立方星技术的迅速发展已经使新型小卫星任务成为可能。"飓风全球导航卫星系统"(CYGNSS)就是一个很好的例子。它所包含的 8 个质量为 28 kg 的航天器并不是立方星,但是使用了商用器件和裁剪的任务保障方法,其最终成本为每个航天器500 万美元(不包括有效载荷);而相比之下,"磁层多尺度任务"(Magnetospheric Multiscale,MMS)[③]的成本为每个航天器 1.65 亿美元(Tumlinson,2014)。CYGNSS的实验卫星被称为"技术验证星-1"(TechDemoSat-1),质量为 157 kg(Foti et al.,2017)。该实验卫星于 2014 年发射,用于验证观测飓风的全球导航卫星系统反射测

---

[①]　美国国家科学院发布的《十年调查》下载地址:http://sites. nationalacademies. org/ssb/ssb_052297。

[②]　下载地址:https://www. nap. edu/catalog/23503/achieving-science-with-cubesats-thinking-inside-the-box。

[③]　译者注:"磁层多尺度探测任务"实现了 4 颗多尺度磁层卫星以最小间距飞行,这也是多颗卫星在轨道上编成的最紧凑编队。任务发射于 2015 年 3 月 3 日,将对空间磁场重联展开前所未有的细致研究。

量技术(Global Navigation Satellite Systems-Reflectometry,GNSS-R),该技术也应用于 CYGNSS。"技术验证星-1"可以被列为小卫星,但是随后的 CYGNSS 卫星充分利用了立方星的新发展,把卫星质量降低了 5 倍,从而使利用一个小卫星星座在合适时间尺度上监测飓风的形成和发展成为可能。

## 1.2.2　近期的科学潜力:"地平线"上的空间科学卫星任务

一些近期的卫星任务和正在研发的卫星任务都采用了小卫星。小卫星任务的规模不一定小,最近已发射了一些采用若干分布式小卫星的任务,并且还将有更多任务在近期发射。

"凌日系外行星勘测卫星"(TESS)和"电离层连接探测器"(Ionospheric Connection Explorer,ICON)[①]是美国发射的最新"探索者计划"卫星,这两个任务都属于"中型探索者任务",造价都超过了 2 亿美元。前者于 2018 年 4 月发射,质量为 362 kg,旨在搜索附近的系外行星;后者于 2019 年 10 月发射,质量为 291 kg,主要研究电离层。最近两次"小型探索者任务"是 2012 年发射的"核光谱望远镜阵列"(Nuclear Spectroscopic Telescope Array,NuSTAR)[②]卫星和 2013 年发射的"界面区成像光谱仪"(IRIS)卫星。作为对天体物理和太阳物理《十年调查》建议的反馈,NASA 最近增加了小型和中型探索者卫星的发射机会。在天体物理领域,"X 射线成像偏振望远镜"(Imaging X-Ray Polarimetry Explorer,XIPE)将于 2020 年发射,旨在研究中子星和黑洞等致密天体的 X 射线产生。2017 年,NASA 遴选出 5 个太阳物理领域的"小型探索者任务"概念,进行下一步任务论证,此外还公布了几个机会性任务的征集,其中有些任务采用了立方星[③]。预计 2020 年将最终完成 1~2 项任务的遴选。

近年来,由 NASA 资助的空间科学立方星的发射数量显著增加,其中 2015 年

---

①　译者注:"凌日系外行星勘测卫星"于 2018 年 4 月 19 日搭乘 SpaceX 公司的猎鹰 9 号火箭成功发射升空。它不仅是即将于 2020 年前后升空的"詹姆斯·韦伯空间望远镜"的得力助手,也被看作 NASA 此前发射的"开普勒空间望远镜"(Kepler)的继承者。"电离层连接探测器"于 2019 年 10 月 14 日发射,利用干涉仪、紫外摄影机和离子速度计的测量来分析电离层的经常性变化。

②　译者注:"核分光望远镜阵列"是一个高能 X 射线天文望远镜,它的首要任务是观测能量范围在 8~80 KeV 的 X 射线源。"太阳界面区成像光谱仪"卫星旨在观测太阳的过渡层,尤其是色球层的物理状态。

③　https://www.nasa.gov/press-release/nasa-selects-proposals-to-study-sun-space-environment。

发射 2 次,2018 年发射 8 次,2019 年计划发射至少 9 次[①]。在新近遴选出的立方星任务中,"地球同步转移轨道卫星"(GTOSat)将成为第一个在地球同步转移轨道(Geostationary Transfer Orbit,GTO)上运行的科学立方星。该立方星将对地球辐射带进行关键观测,卫星采用抗辐射的 6U 平台,可作为未来磁层星座任务的探路者[②]。在地球科学领域,目前在轨的"飓风全球导航卫星系统"(CYGNSS),以及即将发射的"热带"卫星群任务(TROPICS)[③]都将证明星座方法在地球科学领域的应用。特别需要指出的是,由 12 颗立方星组成的 TROPICS 卫星群可提供 30 分钟重访,这对监测迅速发展的风暴系统尤为重要(图 1.6,参见彩图)。未来的近地轨道星座可以利用基于全球定位系统(GPS)的相对定位技术来精确自主地确定编队卫星的相对位置,而这是编队飞行、编队维护以及实现科学目标所必需的(Causa et al.,2018)。

欧洲仍然使用传统小卫星进行空间科学研究,虽然应用比例并不高,但追求重要的科学目标。法国国家空间研究中心(CNES)于 2016 年发射了卫星质量为 330 kg 的"显微镜"(Microscope)卫星,验证了爱因斯坦广义相对论的重要组成部分——等效原理(Equivalence Principle)[④]。Microscope 卫星的目标是将验证精度提高至 $10^{-15}$,其观测精度比迄今地球上进行的实验提高了 100 倍(Touboul et al.,2017)。PROBA-3 是 ESA"星上自主项目"的系列任务之一,预计将于 2020 年发射。该任务包含两个航天器,质量分别为 340 kg 和 200 kg,它们将以 150 m 的飞行间距制造一次人工日食进行日冕研究。此项任务还将验证精确编队飞行[⑤]。"系外行星特征卫星"(CHEOPS)是 ESA 的第一个小型任务(S-class),质量约为 300 kg,其主要使命是研究已知的系外行星的特征。尤其值得注意的是,CHEOPS

---

① NASA 的科学立方星任务数据来自 Larry Kepko 于 2017 年 10 月所做的报告,题目为"SMD 立方星计划新进展"。检索日期为 2019 年 2 月 20 日;下载地址为 https://smd. prod. s3. amazonaws. com/science-red/s3fs-public/atoms /files/ kepko. smallsats-apac_october ％202017. pdf。发射日期来源于 CSLI 网站:https:// www. nasa. gov/content/past-elana-cubesat-Launch。MarCO 发射日期来源于该卫星任务网站:https:// www. jpl. nasa. gov/cubesat/ / MarCO . php。

② https://www.nasa. gov/feature/goddard/2018/nasa-s-new-dellingr-spacecraft-baselined-for-pathfinding-cubesat-mission-to-van-allen-belts。

③ 译者注:"热带"卫星群任务,全称为"基于小卫星星座的降水结构和风暴强度时间分辨观测"(The Time-Resolved Observations of Precipitation Structure and Storm Intensity with a Constellation of Smallsats,TROPICS),它的 12 颗立方星分布在 3 个近地轨道平面上,每颗立方星均可进行高精度的扫描,利用辐射计精准地观测风暴内部的温度、湿度、降水和云特性参数。

④ 译者注:等效原理是广义相对论的基础,目前在地球上验证精度已达 $10^{-13}$。

⑤ http://www. esa. int/Our_Activities/Space_Engineering_Technology/Proba_Missions/ About_Proba-3,检索日期为 2018 年 7 月 8 日。

图 1.6　NASA"热带"(TROPICS)卫星任务概况[图片来源:TROPICS 团队,麻省理工学院林肯实验室(TROPICS 是"利用小卫星星座进行降水结构和风暴强度的时间分辨观测"的英文缩写),卫星网站链接为 https://tropics. ll. mit. edu/cms // Mission-Overview]

将测量系外行星的半径,这些数据与地面天文台测量的质量数据相结合,将首次计算出系外行星的密度。

　　在欧洲,即将发射的立方星的数量似乎正在增长。ESA 用于"一般支持技术计划"的预算为 800 万欧元,用来在 3 年内支持一项或者多项任务,以达到显著改善系统性能或提升新的技术应用的目的。这方面典型的例子是"高能快速模块化组合卫星"(High Energy Rapid Modular Ensemble of Satellites, HERMES)[①]。这是一项意大利高能(KeV~MeV)天体物理任务,此前在这一科学领域只有大型空间任务。HERMES 是一个近地轨道纳卫星星座,每颗纳卫星质量小于 10 kg, X 射线探测器的有效面积大于 50 cm$^2$(能量范围从几千电子伏特到 1 MeV),且具备非常高的时间分辨率。HERMES 的主要科学目标是研究并准确定位高能天体物理现象,如伽马射线暴、引力波的电磁对应体[产生于致密天体合并现象,正如最近"激光干涉引力波天文台"(LIGO)和"欧洲引力波天文台"(VIRGO)所观测到的]以及快速射电暴发(fast radio bursts)的高能对应体。目前,一个由 3 颗卫星组成的技术验证星座正在研发中,预计于 2020 年发射,之后将发射科学验证卫星任务。

---

　　① 　http://hermes. dsf. unica. it/index. hml。

最终的目标是在不同轨道上建立一个由数十颗卫星组成的星座,以提供全天空范围内精确度优于 1 度的瞬态位置。

若干国际合作小卫星任务目前正在研发中。以色列航天局(ISA)和法国国家空间研究中心(CNES)联合研发并发射了"植被和环境监测新型微卫星"(Vegetation and Environment Monitoring on a New Micro-Satellite, Venμs)①。这颗卫星干重达 250 kg,配备一台超光谱相机,可以同时观测 12 个波段。该卫星对全球范围内的科学目标进行重访(频率可达两天一次),目的是研究植被的演变;同时,该卫星对基于霍尔效应推进器(Hall-Effect thrusters)的独特电推进系统(electrical propulsion system)进行在轨验证,此电推进系统使用无毒氙气,可以尽量减少推进剂的使用,同时实现灵活的轨道机动。"星载高光谱陆地和海洋应用任务"(Spaceborne Hyperspectral Applicative Land and Ocean Mission, SHALOM)是以色列航天局和意大利航天局的联合卫星任务,将研制几颗通信和地球观测小卫星,主要用来发现和识别地表、水体及大气中的污染物。

目前,一些国家正在进行低频空间射电干涉测量的概念研究(图 1.7,参见彩图)。"轨道低频阵列"(Orbital Low Frequency Array, OLFAR)概念将在月球轨道部署一个大型星座,包含若干小型航天器(Rotteveel et al. ,2017)。2018 年 5 月 21 日,"荷兰-中国联合低频探测器"(Netherlands Chinese Low Frequency Explorer, NCLE)发射升空,迈出了实现轨道低频阵列卫星任务的第一步(Castelvecchi,2018)。

将来在红外和光学波段也可以进行干涉测量。"可重构的自主装配空间望远镜"(Autonomous Assembly of a Reconfigurable Space Telescope, AAReST)②是一项国际合作卫星任务,美国加州理工学院(CalTech)、喷气推进实验室(JPL)、英国萨里大学(University of Surrey)和印度空间科学与技术研究院(IIST)等四家单位正在进行任务概念研究,旨在研制一个光学望远镜,其"主镜"由安装在分布式立方星上直径为 10 cm 的圆形镜面组成(Sweeting,2018)。

---

① 译者注:"植被和环境监测新型微卫星"于 2017 年 8 月 2 日发射,该卫星是以色列发射的首个专门用于农业与生态研究的卫星,可获得特定位置的高分辨率照片,记录土地状况、积雪程度、植被情况、造林进程、农业、水源质量等领域的资料,从而跟踪荒漠化、侵蚀、污染、自然灾害等与气候变化相关的现象。卫星能够在相同的光照条件(即太阳位置相同)下重复拍摄相同地点不同时期的照片,从而更准确地跟踪变化中的环境问题。

② 译者注:"可重构的自主装配空间望远镜"旨在通过 2 颗 3U 立方星(子星)与 1 颗 9U 纳星(母星)进行自主分离和重组,来验证空间望远镜的自主装配和组合技术。中央纳星上装有 2 个固定反射镜和 1 个悬臂焦点平面组件,2 颗 3U 立方星均携带电动自适应反射镜。

**图1.7** 目前,一些空间干涉测量的任务概念正在论证中,其中包括本图所示的 NASA "日出"任务概念(SunRISE mission concept),该任务将采用 6U 立方星组成的 小星座来测量太阳射电暴(solar radio bursts)(图片来源:"日出"任务概念研 究团队)

近年来,微纳卫星也开始利用搭载发射机会,从近地轨道向深空探索领域进 军。与典型的使用具有高技术成熟度的部件并搭载一整套有效载荷的深空探测任 务相比,这些微纳卫星任务主要用于科学调查或测试新技术。

2014 年搭载日本小行星探测器"隼鸟 2 号"发射升空的"光学导航临近天体近 距离飞越卫星"(PROCYON)①是第一个深空探测微卫星,其发射质量为 67 kg,同 时该卫星也是第一个由大学主导的深空任务。该卫星摆脱了地球引力进行远距离 飞越,并于一年后返回。PROCYON 任务验证了全功能卫星平台,包括低、中、高 增益天线,以及动量轮和冷气喷气、电推进系统、望远镜和相机。PROCYON 任务 从建议书、研发到发射,共历时 14 个月,并且大部分团队成员都是东京大学的学 生。"地月拉格朗日平动点的 6U 航天器"(EQUilibriUm Lunar-Earth point 6U Spacecraft,EQUULEUS)和"利用半硬质纳米撞击器的杰出月球探测技术验证" (Outstanding Moon exploration Technologies demonstrated by Nano Semi-Hard Impactor,OMOTENASHI)是两个后续任务,均为 JAXA 和东京大学研发的 6U

---

① 译者注:2014 年发射的世界上最小的探测器"PROCYON"成功观测到彗星,并测定了 其释放的水蒸气量。据悉,重量 100 kg 以下的超小型探测器取得这一成果在世界上尚属首次。

立方星任务。EQUULEUS 将是第一颗采用水工质电热式推力器到达月球拉格朗日点的立方星。OMOTENASHI 卫星将成为最小的月球着陆器。最近,国际空间站释放了"带有 Gossamer 减速伞和 GPS/铱星通信系统的再入卫星"(re-Entry satellite with Gossamer aeroshell and Gps/Iridium,EGG),这种减速伞将来可能用于再入大气层或进入轨道。

2 颗"火星立方 1 号"(Mars Cube One,MarCO)最近完成了火星之旅,为"洞察号"(InSight)着陆器的入轨、下降和着陆提供数据中继。曾被命名为 AIM/AIDA 的 ESA "赫拉(Hera)任务"[①],其航天器(母星)携带 2 颗立方星,并将在目标小行星迪迪莫斯的"小月亮"附近释放(Perez et al.,2018)[②]。下一步技术验证项目将着眼于实现立方星进军深空探测,其中包括"相关环境中的行星际纳卫星探路者"(Interplanetary Nano-Spacecraft Pathfinder in Relevant Environment,INSPIRE)(Klesh et al.,2013)和"深空内部扫描立方星"(Deep Interior Scanning Cube-Sat,DISCUS)(Bambach et al.,2018)。这两颗立方星将配备一台矢量磁强计(vector magnetometer)和一台成像仪。得益于"探索任务 1 号"(Exploration Mission 1)的资助和美国"空间发射系统"(Space Launch System,SLS)火箭的首飞,另有 13 颗立方星已准备就绪,随时可以飞往月球和更远的深空[③]。

在未来,不同系列的航天器可以为不同类型的卫星任务提供各种选项。这些航天器从高技术芯片卫星,如"阿卫星"(Attosat)或"飞卫星"(Femtosat),到传统的基于小型化空间技术增强能力的大型航天器等,不一而足。搭载小卫星探测器的大型航天器可以在太阳系范围内运行。它们既可以作为主要任务的一部分,也可以作为纯搭载卫星。

---

① 译者注:国际天文学联合会于 2020 年 6 月 23 日正式将迪迪莫斯(Didymos)小行星的"小月亮"命名为"迪莫弗斯"(Dimoyphos)。"迪迪莫斯"和"迪莫弗斯"共同组成了一个双小行星系统。2021 年 7 月,NASA 将发射"双小行星重定向测试任务"(Double Asteroid Redirection Test,DART)(这项任务也被形象地称为"飞镖任务")。这将是人类第一次全尺度验证小行星偏转技术,其目标小行星是"迪莫弗斯"。欧洲空间局(ESA)将于 2024 年随之发射"赫拉任务"(Hera)。由 DART 任务和赫拉任务组成的国际合作"小行星撞击和偏转评估任务"(Asteroid Impact and Deflection Assessment,AIDA)将对能够用来保护地球免于危险小行星撞击的偏转技术进行演示验证,偏转技术将使小行星变轨,偏离撞击地球的轨道。

② https://phys.org/news/2019-01-cubesats-hera-mission-asteroid.html。

③ NASA 的新空间发射系统(SLS)。

## 1.2.3　局限性和技术挑战

美国国家科学院于 2016 年发布了关于立方星的报告,对空间科学进步所需的技术和最近的技术进展进行了概述。报告特别突出了推进系统、通信、传感器小型化、抗辐射部件和亚角秒姿态控制领域的进展。美国防务分析研究所(IDA)在关于小卫星的报告(Lal et al.,2017)中,对小卫星的技术趋势进行了整体评估,涉及高带宽通信和在轨处理,以及小型化进展、轨道碎片监视(Lupo et al.,2018;Santoni et al.,2018)和碎片移除技术。很多技术是受商业市场驱动而发展起来的,也是空间科学卫星任务所需要的。目前的技术发展趋势解决了一些近地轨道小卫星的局限性,其中包括:

(1) 降低小型化元器件的噪声:基于滤波技术的软件方案降低了对噪声的敏感性。

(2) 姿态和轨道控制能力:最近提出的小型化反作用轮技术实现了在低功耗情况下提高姿态控制,并为电推进系统提供了轨道控制。因此,即便是 1U 的立方星,也能实现提高仪器指向精度和提升编队能力,例如下面三个小卫星:"光通信与传感器演示验证"(OCSD)小卫星①、"伍兹堡大学实验卫星-4"(UWE-4)小卫星和"远程通信地球观测"(TOM)小卫星。科拉格罗西(Colagrossi)和拉瓦尼亚(Lavagna)于 2018 年提出了一种全磁姿态控制子系统(Colagrossi,Lavagna,2018)。

(3) 通信链路容量:光学链路技术的新进展使链路容量在晴空条件下可以超过 100 MB/s(如 OCSD 任务、QUBE 任务和 TOM 任务),此外,超小型 X 波段收发器也已具备应用条件(如 MarCO 任务)。

(4) 卫星延寿:先进的故障检测、识别和恢复(FDIR)以及备份概念(redundancy concepts)保证了在轨卫星的预期寿命(即使采用了商用货架组件)。[例如,UWE-3 已连续运行超过 3 年,尽管期间遭遇了单粒子翻转事件(SEU)和单粒子锁定事件(SEL)。]

(5) 地面段:目前已经建立的若干个大学地面站网络,如全球卫星运控教育网络(GENSO)和大学空间工程联盟(UNISEC),支持小卫星数据的频繁传输,以减

---

① https://aerospace.org/story/communicating-and-converging-cubesats,检索日期 2019 年 2 月 22 日。

轻在轨数据储存和处理的压力。康斯伯格卫星服务公司的商业网络正在为小卫星提供全球覆盖服务。

到目前为止,空间科学小卫星,尤其是立方星任务的数据返回受限于地面站的可用性(即受限于任务成本)。随着空间科学和商业两用大型星座任务的发展,卫星通信可能会变得更加困难。包含数以百计或千计卫星的星座任务,尤其是具有成像能力的地球科学星座任务,将产生大量数据。斯威廷在2018年指出:"未来十年,从小卫星下行的累积数据量将达到3.9艾字节(exabytes)(1艾字节 = $10^{12}$ MB)。传统的射频能力很可能满足不了这种需求……"目前,已有几家研发单位正在攻关用于太空的低功率光学终端,数据下行速率将高达10 Gb/s(Sweeting,2018)。无论如何,星上数据处理还是必要的,它可以尽量减少传输到地面的数据量。对未来的空间科学任务来说,先进的数据处理技术(如人工智能)是有益的,也是必要的。商业星座和卫星运控系统也可能需要接近实时可用的数据分发系统。这样的系统会为空间科学任务创造新的机遇。

> **研究发现1.7** 随着科学技术的进步,未来对进一步实现编队飞行、星间通信、数据压缩和巨型星座部署技术的需求将会随之增强。

深空探测任务还面临着独特的挑战。前文所引用的报告中没有覆盖深空探测任务所面临的挑战,因此将在本节进行更详细的讨论。

### 1. 远程通信

目前,深空探测立方星的设计是为了执行低数据量的测量,这限制了科学观测。此外,深空探测任务需要X或Ka波段的地面天线,因此需要大型空间机构地面站网络的支持。过去的做法是,小卫星任务所需地面站时间需要通过协商和谈判来解决(取决于地面站的机会窗口,优先权也比其他卫星任务低),并且通常只能一次协商一颗或两颗卫星的地面站时间。深空探测的"民主化"或"大众化"要求对数十个甚至数百个纳卫星进行支持,尤其是在发射和早期运控阶段。例如,和"探索任务1号"(Exploration Mission 1,EM-1)一同发射的13颗立方星中的大多数将在释放后的两天内执行一次关键的机动,为此卫星需要下行和上行链路来进行卫星运控和精确的测定轨(多普勒双向测距)。目前,光链路设备和X波段收发器领域技术的发展有希望为上述问题提供解决方案。日益增多的分布在全球的小型地面站将在未来提供连续覆盖,这类似于无线电爱好者在超高频/甚高频段中支持立方体卫星团队的超高频地面站网络所起到的作用。

### 2. 能源

因为距离太阳太远,发电需要大型太阳能电池阵列,如"罗塞塔"(ROSETTA)

彗星探测任务,否则就需要使用替代能源。核能发电机已经用在星际航天器上(如"卡西尼"任务和"伽利略"任务),但是现在还无法用于立方星。因此造成的结果是,立方星任务只能使用非常有限的电池存储能源,这就要求卫星运控必须十分小心,不能浪费这些有限的能源。

### 3. 推进

到目前为止,大多数立方星的推进系统都是冷气推进系统或热气推进系统,部分原因是它们的成本和复杂性较低。目前,多家公司正在研发一大批用于立方星和微卫星的各种类型的电推进系统,这些推进系统的技术成熟度(TRL)在 5~7[①]。

AeroCube 8 立方星任务在近地轨道已经验证了微型电推进系统的能力[②]。对于深空探测任务,必须增加总冲量,并辅之以更大的燃料储存能力。像太阳帆[③]这样的创新技术可能会提供解决方案。

### 4. 任务设计

对小型深空探测航天器来说,轨道动力学和导航尤其具有挑战性,这些小型航天器的轨道控制能力有限,但需要到达与大型航天器类似的目的地。和大型卫星相比,小卫星的任务设计同样重要和复杂,有时甚至更加重要,因为小卫星更加依赖于专家的人力投入和先进的设计工具。因此,小卫星的任务设计大多由航天机构完成。对于开源的且不涉及国际武器贸易条例(ITAR)的卫星任务研发的支持将降低深空纳卫星的成本,并鼓励新的利益相关者参与。

### 5. 卫星运控

和任务设计一样,小卫星的运控与大型卫星任务同样复杂和昂贵。大部分深空探测任务的运控工作都是通过"人工监控"的方式执行的。增强卫星自主性将降低任务成本,但由于潜在的风险,目前卫星自主性还未在高造价的任务中充分发挥潜力。目前正在为集群和编队飞行卫星任务开发近地轨道的自动化运行能力,而深空探测任务将受益于此。但是由于可用地面站有限,深空探测纳卫星的自主度要求更高。对自主运行操作和导航技术发展的支持,将使小卫星可以用于深空探测,并最终降低大型任务成本。

---

① https://sst-soa.arc.nasa.gov/04-propulsion。

② http://spl.mit.edu/news/aerocube-8-cd-launch-mit-spls-electrospray-propulsion-space。

③ 译者注:太阳帆(solar sails)是利用太阳光的光压进行宇宙航行的一种航天器。由于光压所产生的推力很小,所以航天器不能从地面起飞,但在没有空气阻力的太空,这种小小的推力能为具有足够帆面面积的太阳帆提供所需的加速度。

### 6. 发射机会

要增加空间科学卫星的发射机会也需要航天发射活动的增强。如1.1.3节所述,近地轨道卫星任务的发射机会已经显著增多。目前正在开发的小卫星发射火箭(见表4.1防务分析研究所报告,Lal et al.,2017)拓展了立方星的轨道范围,使其不再受运载火箭的限制。例如,2018年12月,火箭实验室(Rocket Labs)的"电子号"(Electron)火箭将13颗立方星送入近地轨道①。但是对深空探测任务来说,搭载发射的机会比较稀缺(图1.8,参见彩图)。2015年,"猎鹰9号"运载火箭将美国国家大气海洋局(NOAA)的"深空气候观测平台卫星"(Deep Space Climate Observatory,DSCOVR)送入地月第一拉格朗日点,未使用的火箭运力为2500 kg;NASA的"凌日系外行星勘测卫星"(TESS)被送入地月转移轨道,剩余火箭运力为3000 kg②。对小卫星定期开放月球、拉格朗日点或行星际任务剩余火箭运力,将改变深空探测立方星的发展格局。这样的机会可能很快就会变成现实:NASA最近承诺每次空间科学任务发射都会配置搭载有效载荷适配器,为小航天器预留座位③。

> **研究发现1.8** 重大的技术进步为小卫星开辟了新的机会,解决了一直以来限制其科学回报的各种问题。然而,对深空探测小卫星来说,还有一些其他的特殊挑战。

总之,立方星的出现正在推动技术的重大进步,这些技术进一步推动了像"飓风全球导航卫星系统"(CYGNSS)一样的其他小卫星任务(非立方星)的发展。尽管立方星任务有很高的科学前景,但是2017年用于科学探索的立方星任务数量不到总立方星任务的12%,使用较大小卫星的空间科学探测任务次数甚至更少。尽管如此,立方星或采用立方星技术的小卫星已经开始出现在大型商业卫星星座中。虽然由几十颗或更多的纳卫星组成的星座在空间科学方面的科学潜力还有待验证,但是由立方星和较大小卫星组成的或两者协同工作的空间科学卫星任务具有巨大的科学前景。

---

① http://rocketlabusa.com/news/updates/rocket-lab-successfully-launches-nasa-cubesats-toorbit-nfirst-vent-class-launchser-vices-mission/,检索日期2019年2月22日。

② https://spacenews.com/government-agencies-prepare-for-piggybackflights-secondary-payloads/,检索日期2019年2月22日。

③ https://spacenews.com/nasa-bolsters-smallsat-science-programs/,检索日期2019年6月8日。

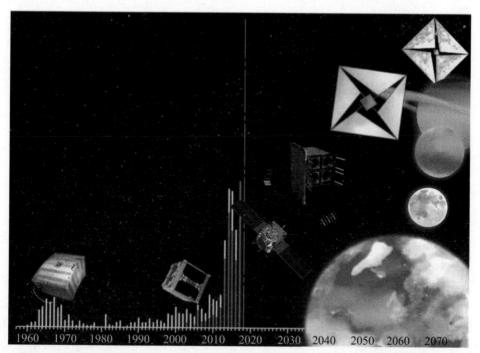

图 1.8　浅色条状图显示了从 1957 年到目前立方星的演变(基于 Swartwout 的数据)。深色表示立方星的数量。随着"火星立方任务"完成了火星之旅、SLS-EM-1 任务和 Hera 任务即将发射并计划对迪迪莫斯双小行星系统进行探测,"搭载"的发射模式已经超出了近地轨道。上述 3 项卫星任务都采用了立方星模式。未来的卫星任务可能要求航天器的质量更低,因此需要新的航天器概念,如由深空探测任务分离舱部署出来的立方星集群或者激光推进的超小型芯片卫星⌊Katrine Grønlund 艺术假想图:Oscar-1 卫星艺术图获得授权使用:http://www.arrl.org/space-communication;DTUsat-1 卫星图像由 DTUsat-1 卫星团队提供;Hera 卫星任务图片来源:https://www.esa.int/our_activities/space_engineering_technology/asteroid_impact_mission/asteroid_impact_deflection_assessment_mission;国际空间站四星释放器立方星图像获得授权使用:https://www.isispace.nl/product/quadpack-cubesat-deployer/⌋

# 第 2 章
## 小卫星未来展望

本章主要从"我们的小卫星社区"(小卫星的昨天和今天)转向介绍代表小卫星未来愿景的高峰(小卫星的明天),即以目前(甚至在未来几十年)的技术水平尚无法实现的开发活动或卫星任务。其中,2.1 节讨论地球科学和地球空间科学领域未来的发展。在这两个领域中,由相互通信并与地面通信的成百上千的小卫星构成的全球系统不仅将对空间科学产生巨大影响,而且对社会也会产生巨大影响(后者超出了本书的讨论范围)。2.2 节探讨发射小卫星集群对太阳系天体进行探测的可能性,例如对 2061 年返回内太阳系的彗星 1P/哈雷(comet 1P/Halley)进行探测。2.3 节描述由小卫星组成的合成孔径光学望远镜(synthetic aperture optical telescope)的潜力,其可能应用包括对太阳以外的恒星进行成像。最后,2.4 节讨论最近引起公众广泛关注的"突破摄星"计划(Breakthrough Starshot)的星际任务的可能性。本章列出 4 个愿景的目的,并不是追求全面和权威,只是作为有可能实现的例子来预测小卫星未来几十年的潜力。

## 2.1　小卫星在地球科学和地球空间科学领域中的潜力

立方星和小卫星有可能在地球观测、太阳和空间物理等科学领域中做出独特的贡献。尤其是小卫星将使大型星座任务成为可能,从而为在太空进行科学研究提供新工具。在地球科学领域的应用包括但不限于地表成像、气象应用、污染研究和太阳辐照度测量等。在空间物理领域,美国最近的《十年调查》突出强调了使用

"系统-科学方法"(systems-science approach)研究日地系统的重要性。美国国家科学工程和医学院(NASEM)于 2016 年发布的关于立方星的报告强调了多点测量对于实现"系统-科学方法"的重要性,并指出多点观测是小卫星的一项主要优势。

　　本节将简要描述几项任务概念,正是由于小卫星所带来的机会使它们变得更容易实现。这些任务概念的想法并不是新创举(Esper et al.,2003);它们所面临的挑战不在于所需要的技术,而更多的是现有可支配的预算。商业航天的发展(图 2.1,参见彩图)可能为降低成本和实现这些任务提供途径。

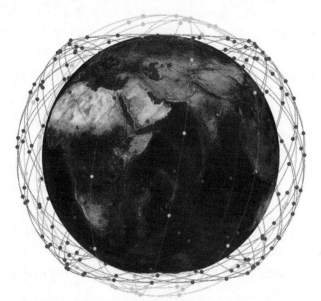

**图 2.1　大型商业卫星星座示意图[图片来源:Telesat(加拿大电信卫星公司)]**

## 2.1.1　地球科学超级星座任务

　　在地球科学领域,具有大量近地轨道大型星座任务(由成百上千颗卫星组成)的项目申请。由 5～10 颗小卫星组成的小型星座任务已经在轨运行,如第 1 章所述的"飓风全球导航卫星系统"(CYGNSS)(Ruf et al.,2018)。从应用来看,具体任务的差异性主要体现在空间分辨率、时间分辨率、光谱分辨率等方面。但是总体来说,使用小卫星(和大型地球观测卫星相似)监测地球应该尽可能地提供更高频率、更准确、更精确和更完整(包括波长、偏振等)的更多数据。

　　利用小卫星进行地球监测的主要优点如下:

(1) 在星座任务中采用大量小卫星可以增加卫星重访频率,以便研究短时变化,也可以用于卫星追踪。

(2) 由于标准化和小型化,大规模小卫星生产可以降低成本。

(3) 由于发射成本与卫星质量成正比,所以小卫星的发射成本更低,即使是小卫星星座任务也要比单一大型卫星任务便宜。

(4) 小卫星可用于验证大型卫星新任务概念的可行性。

然而,上述所有优势都不能与物理定律相左(如分辨率),小卫星研发要遵守这些物理定律。用于大型卫星的技术应该针对小卫星的自身限制做出适应性修改,这些限制主要包括小卫星的尺寸和质量。为了弥补小卫星尺寸上的不足,可以采用更高的工作频率或者人工增大孔径(D)来提高空间分辨率。例如,合成孔径雷达和干涉测量等技术可以用于小卫星星座任务。虽然这两项应用都面临技术挑战(如高频接收机、发射机和控制及稳定、推进技术),但是未来的发展不应该把如此有前景的技术进步排除在外。

小卫星的另一项重要潜力是可以与大型卫星任务协同工作,从而补充大型卫星的能力。"陆地卫星"(Landsat)和"哥白尼/哨兵 2 号"(Copernicus/Sentinel-2)已经使用了这种方法,并结合了"行星星座"的数据,因此每天产生的高分辨率图像可以对精确的多光谱数据进行补充。另一个例子是 ESA 的地球探测任务——"荧光探测器"(Fluorescent Explorer,FLEX)。按照设计,它将与"哥白尼/哨兵 3 号"一起飞行。这种任务之间的协同可以用来开发新技术,确保不同任务之间测量数据交叉定标,也可以确保科学应用的数据质量。

一些小卫星的创新概念正在研究中。例如,与较大卫星一起编队飞行的小卫星,其中发射机安装在大卫星上,小卫星只安装接收雷达天线,以实现双站测量。目前,处于 ESA 卫星设计阶段的 L 波段"阿根廷微波观测卫星"(Satélite Argentino de Observación COn Microondas,SAOCOM)的伴星(Companion Satellite,CS),证明了小型被动合成孔径雷达(Synthetic Aperture Radar,SAR)可以显著提升主任务的科学目标,该伴星与 SAOCOM 相伴飞行。如果使用得当,小卫星的灵活性和一些相关技术可能会产生很多其他新技术应用(目前有些新应用已经落地),包括视频能力、实时成像和实现用户在地面直接向有效载荷发送指令等。

值得注意的是,我们不应该仅仅考虑高重访率和(空间)分辨率之间的对立关系,得出对用户而言重访率比分辨率更重要的结论(虽然通常是这样,但并不绝对)。实际上,光谱/辐射参数也可以促进一些关键应用,这些应用适用于大气观测或水圈观测,甚至适用于"经典"陆地成像。也就是说,实际的原理和反演可以通过多点测量完成,甚至可以使用不同技术,如光学和合成孔径雷达等。这样可以捕捉到不同周期之间的相互作用信息,这是思路上的转变,而不是简单的在空间或时间分辨率上的改进。

为了实现未来任务,空间科学可以抓住机遇,充分利用商业航天领域的新发展。正如第 1 章中所讨论的,商业航天对地球观测小卫星的兴趣正在迅速增长。例如,在 2017 年发射的 328 颗小卫星中,有 103 颗是同时发射的(2017 年 2 月 14 日由印度空间研究组织的极地轨道运载火箭送入太空,其中包括"行星实验室"的小卫星)。在 328 颗卫星中,三分之二的卫星用于对地观测,其中质量在 10 kg 以下的占 89%,10~100 kg 的占 7%,100~500 kg 的占 4%;剩下三分之一的卫星用于技术及科学应用(31%)和通信(2%)(数据来源:CATAPULT,2017)。值得注意的是,328 颗卫星的总质量低于 2002 年 ESA 发射的一颗环境卫星(ENVISAT)的质量,该卫星重 8 吨,搭载了 10 种不同仪器①。

空间科学星座任务可以充分利用技术的发展(标准化和小型化),并学习工业制造、集成和测试方法。商业航天的新发展也带来了购买商业数据或将科学仪器放到商业有效载荷平台上的机会。尽管到目前为止,确实已经有了成功的例子,但是这种伙伴关系仍然还面临挑战。

不管采用何种模式,都应考虑小卫星获取数据的一些关键因素,其中包括:

(1) 数据的可持续性:如何从科学的角度确保这些卫星任务获取的数据是有长期承诺的,而这正是科学用户所需要的。

(2) 数据政策:所有数据都尽可能地完全免费开放是好的科学发展计划的先决条件,应该考虑如何确保这样的数据政策②。

(3) 商业和科学需求之间潜在的数据获取冲突:当前很多小卫星的进步都是由商业实体驱动的,这些商业实体提供了大量的地球轨道卫星数据集。但是这些数据用于科学目的时,要确保此类商业卫星任务的科学潜力不会受到商业利益的妨碍。

---

① 有趣的是,ENVISAT 卫星上的有效载荷成本为每千克 30 万美元,这个成本与 2017 年发射的典型的单一有效载荷小卫星相比还是非常有竞争力的。同时,ENVISAT 卫星自 2002 年至 2012 年持续在轨运行了 10 年,是其标称设计寿命的两倍。

② https://earth. esa. int/web/guest/-/-esa-earth-observa-data-policy-7098,2010,或欧盟委员会授权条例第 1159/2013 号,https://eur-lex. europa. eu/legcontent/en/txt/html/? uri = CELEX:32013 r1159&from = EN。

（4）兼容性：小卫星应该被当作大生态系统中的一个元素，通常它们是为了助力"哥白尼哨兵"（Copernicus Sentinels）这样的机构卫星，以加强其观测能力。

（5）数据下行：大量数据需在星上进行数据处理，这样可以只下行有用数据（如星上数据处理可以过滤云层笼罩的不清晰图像）。

（6）数据利用：数据传输到地面后，小卫星任务面临的挑战更多的是大量数据的处理和利用（如大数据、人工智能及卫星数据和非卫星数据等不同类型数据的融合），它们比卫星研发本身更重要。

进一步讲，空间科学小卫星的研发总体上应该服务于用户群体的需求，以此来避免"技术推动"的方式（technology push approach），因为这种方式可能会产生大量未经定标和无用的数据。而且也能避免夸大小卫星的能力，如对小卫星能力的承诺无法兑现，这将不利于空间科学界对小卫星的进一步利用。开放的数据政策和数据国际共享非常重要，这一点怎么强调都不过分。世界各地的生物学家为收集和分享迁徙动物数据而建立了"移动银行计划"数据库，这个数据库可以作为范例，对小卫星数据共享计划也会有启发①。

## 2.1.2 磁层星座任务

小卫星在磁层研究中的重要作用已得到广泛认可（Shawhan，1990）。数十年来，空间物理界一直在讨论大型磁层星座任务的必要性（Angelopoulos，Panetta，1998；Fennell et al.，2000）。由3～5个航天器组成的小型星座已经使空间科学焕然一新，如ESA的"星簇计划"（Cluster）和"集群"卫星（SWARM），美国的"亚暴期间的事件演化过程及其大尺度相互作用任务"（THEMIS）和"磁层多尺度任务"（MMS）。其中在小卫星的使用方面，THEMIS起到了良好的探路者作用；该任务中5颗77 kg（干重）的卫星由一枚"德尔塔II"（Delta II）火箭以一箭五星的方式发射升空，并且已经在轨运行了12年。但是，从由5颗卫星组成的星座任务到由几十颗甚至几百颗空间科学卫星组成的星座任务的跨越至今还未实现。

磁层、电离层和热层是一个耦合系统，反映了上方太阳风和下方大气的驱动作用。若要理解太阳风的能量是如何耦合进入这个系统以及与相邻空间区域之间的相互作用，需要在广阔的区域进行多点测量。与分布在全球气象站或分布在广阔海洋中的观测浮标一样，对整个磁层系统的分布式测量数据和复杂的计算模型

---

① https://www.movebank.org/。

（图 2.2，参见彩图）（Spence et al.,2004），将使我们有能力理解和预测空间环境。

**图 2.2　在 NASA 的"磁层星座"(MagCon)任务概念中,36 个航天器叠加在磁层流体动力学(MHD)模拟的地球磁层上[转自 Spence 等人(2004)]**

从目前来看,要把这样一个大型磁层星座任务概念付诸实践仍需几十年。然而,小卫星的快速发展让我们有理由对未来保持乐观。小型化仪器的研发正在进行中。例如,由 NASA 戈达德空间飞行中心研制的 6U"德林杰"(Dellingr)立方星为小型的科学磁强计和质谱仪提供了测试平台;美国科罗拉多大学研制的"科罗拉多学生空间天气实验"(Colorado Student Space Weather Experiment,CSSWE)立方星上搭载了"相对论电子质子望远镜"(REPT),这台高能粒子探测器是"范艾伦探测器"(Van Allen Probes)的小型化版本。目前,戈达德团队正在研究一种基于 Dellingr 立方星设计的抗辐射卫星平台"GTOSat"。接下来的关键一步将是如何制造和测试大批量的同型号卫星,而空间科学界在这方面需向工业界学习。

利用商业航天器上的空间来搭载大型星座任务的某些功能仪器也是可能的。例如,在商业近地轨道卫星上搭载有效载荷可以对电离层边界进行特定测量。AMPERE 项目就是一个现有的例子,该项目由 NSF 资助,使用铱星系统(Iridium satellites)上的"姿态测定与控制系统"(ADCS)磁强计来探测极光区域的场向电流(Anderson et al.,2000)。这个项目是通过三方合作来实现的:商业航天工业部门、高校研究人员和 NSF(代表政府)。最近的一项研究使用了行星实验室公司(Planet Labs Inc.)的立方星星座磁强计(Parham et al.,2019)。近期发射的"边缘与盘面全球尺度观测任务"(Global-scale Observations of the Limb and Disk mission,GOLD)提供了另外一种视角,该任务从地球同步轨道对高层大气进行测量。GOLD 上的紫外光谱仪（一种科学仪器）还搭载在美国一颗商业通信卫星 SES-14 上，这颗通信卫星是空客公司（Airbus）为"SES 政府解决方案"公司（SES Government Solutions）研制的。

国际合作为实现诸如磁层星座这样雄心勃勃的设想提供了另一种途径。事实上，1.1 节中所讨论的 QB50 项目中一些航天器安装了等离子体探测仪（朗缪尔探针）。尽管该项目的目标不是科学研究，但它可以作为一种国际合作模式，使大量小卫星的制造和发射更协调有序（另见 3.5 节）。

## 2.1.3　结论和发现

小卫星能够在各个层面（产业上下游、国家和国际层面）促进新的空间科学、应用和商业发展，尤其是通过星座任务和多卫星编队等小卫星任务可以充分利用现成的卫星平台和较短的研发周期等优势。然而，仅仅使用小卫星还无法以更低的成本、更快的速度及更好的效果实现地球科学领域的发展，应将小卫星视为大观测生态系统的组成部分。

小卫星作为新一代卫星为更好地理解地球提供了难得的探索机遇，包括解决观测空白和提供更高频率的测量。在研究小卫星这个极具潜力的新领域发展时，免费、完整和开放的数据政策、有用的标定数据的生成和确保长期可持续数据的获得等相关核心问题要纳入考虑范围。

> **研究发现 2.1**　利用大型卫星星座任务实现地球和空间科学革新发展的机遇是存在的。这一愿景可以通过独立的空间科学任务来实现，也可以通过与在轨小卫星使用数量日益增加的航天工业界合作来实现。空间科学界将从这些大量的小卫星任务获取的数据中受益，前提是这些数据遵循免费、完整和开放的数据政策而服务于科学研究。

## 2.2　太阳系小卫星集群探测

本节将详细介绍一项具有高影响力的行星科学任务概念，并额外介绍两个科学应用的例子，这些科学应用受益于立方星或小卫星组成的大型星座任务（不论组网与否）。

## 2.2.1 "一生只有一次"的行星探测任务

这里所指的行星天体是穿越太阳系的周期非常长的行星天体,即长周期天体(Long Period Object,LPO)。对我们来说,一生只有一次机会能经历这些天体靠近地球。这些天体包括奥尔特云彗星(200 年以上)、无尾彗星[①]以及星际天体(InterStellar Object,ISO),如最近发现的星际天体"奥陌陌"[②]('Oumuamua)(Meech et al.,2017)。对太阳系来说,这不是第一次,也肯定不是最后一次迎来星际访客。长周期彗星是太阳系形成早期最原始的目击者。星际访客被认为是经历了灾难性碰撞的系外行星的抛出物(ejecta of extrasolar planets)。因此,对这些天体进行探索的科学价值是无限的。尤其是,最近的一项研究表明,这些碰撞可能为系外行星系统之间的生命有机体转移提供了一种方式(Berera,2017)。这一发现对理解人类在宇宙中所处的位置和对系外行星进行取样的前景具有重大启发意义。

对长周期无尾彗星和星际天体的观测需要非常广泛的测量范围,包括基本物理特征(形状、密度、形态、动力学特性)、成分属性(元素组成、矿物性质以及至少包括氢、氧、氮、碳的同位素)、地球物理/内部特性(孔隙、内聚力、磁场)、揭示天体起源和可能长期演化的地质特征,以及与太阳风的交互作用,尤其是彗发(coma)存续期间与太阳风的交互作用。实现这些测量的足够小的仪器已经存在了,但是运行它们会有挑战性。具体挑战如下:① 尘埃光谱仪(测量矿物性质、尘埃彗发密度),当与高速物质相互作用时,仪器仍然可以运行;② 原位遥感仪器,如亚毫米波光谱仪(就像"罗塞塔"轨道器搭载的微波设备)可以在允许的安全距离内对挥发物的同位素性质进行测量。其他原位遥感仪器包括彩色成像仪和宽波段光谱仪。进行元素测量则要更加复杂,因为它要求在一定持续的收集时间内与目标天体进行密切的交互作用。在一定程度上,元素丰度的获取可以通过测量目标彗星的彗发/尘埃与太阳风之间相互作用产生的等离子体来实现。

很多因素对长周期天体的探测形成了挑战:① 轨道特性不详,造成没有足够的准备时间来研发一项探测任务;② 倾角范围太大;③ 相遇速度超过 50 km/s,相

---

① 译者注:英文为 Manx object,直译为"马恩岛猫天体",因为这种彗星的形状很像马恩岛上的无尾猫。

② 译者注:"奥陌陌"('Oumuamua)于 2017 年 10 月 19 日被科学家们发现,是已知的第一颗经过太阳系的星际天体。'Oumuamua 在夏威夷语中的意思为侦察兵或信使。'Oumuamua 直径在百米级,大约以每秒 26 km 的速度从天琴座方向冲进太阳系,近乎与黄道面垂直。

遇时间发生在电光火石之间,非常短暂;④ 长周期天体的地球物理活动可能非常活跃,也可能是由多个共轨天体组成的。唯一一次的尝试是探测长周期天体(约75年)1986年与哈雷彗星的相遇。哈雷彗星的近距离飞越是公认的重要事件,当时有不同航天机构研制的6个航天器发射升空;分别是:NASA的"国际彗星探测器"(International Cometary Explorer,ICE)、俄罗斯航天局的"织女星1"(VEGA 1)和"织女星2"(VEGA 2)(Roskosmos)、日本宇宙科学研究所的"彗星号"(Suisei)和"先驱者号"(Sakigake)卫星(日本第一项空间科学卫星任务),以及ESA的"乔托号"(Giotto)探测器。其中三个航天器以前曾研制过,而其余三个航天器都是第一次研制。这些卫星任务之间的协调是由国际航天局顾问小组(Inter-Agency Consultative Group,IACG)执行的,而该小组的成立正是服务于此。因为NASA的ICE处于激波前沿之外,所以实际上它并未与哈雷彗星相遇,但是它在为其他卫星指引方向上起到了重要作用。

实施一项长周期天体或哈雷彗星(2061年回归)的探测任务是具有挑战性的。但可以通过由多个航天机构在有序协调的前提下发射大量的航天器来应对这个挑战(图2.3,参见彩图)。原因也十分简单,一方面,就目前各国的预算来看,由单一航天局发射大量具备多种探测能力的航天器实在难以完成(这些探测能力需要与从这些天体上预期获得的广泛科学知识相匹配)。从另一方面来说,长周期天体和"奥陌陌"在全球引发了巨大热议,这表明通过国际合作来协调对这些天体的探测是非常重要的,也是现实的。

小卫星星座、编队和集群已经被认为是促进新空间科学发展的规则改变者(NASEM,2016)。近年来,小卫星在技术成熟水平上实现了巨大的发展(NASA,2015)。小卫星具有很多优势,尤其体现在先进分布式航天器结构(advanced distributed spacecraft architectures)上,这种结构可以用来应对上述挑战,并有能力在较短的观测窗口实现有益的科学探测。其优势包括:① 通过松散的协调来合成单个大型虚拟仪器(Bandyopadhyay et al.,2016);② 创新的分布式结构、异构测量和数据分析技术;③ 自主运行;④ 通信中继策略;⑤ 星座任务或高效卫星群的新轨道组织方法,有助于实现多点优势探测。

目前,人类还不知道如何靠近速度超过50 km/s的天体。针对哈雷彗星的几项探测任务,虽然很有魄力,但与投入的资源相比,科学上的回报并不多,其原因在于恶劣的重尘埃环境损坏了一些探测仪器。然而,这些任务刺激了欧洲小型仪器的发展,并开启了日本空间科学任务的序幕,因此这次任务是具有里程碑意义的。同样,我们也希望对系外行星碎片和太阳系形成时遗留的最原始的天体等的探测,能够促进空间探测新方法的产生,也希望能够促进各国航天局之间的合作。这类

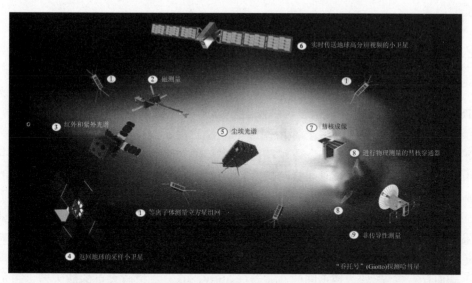

图 2.3　利用分离的星座来探测再次回归的哈雷彗星(2061 年)。本图展示了各种类型的探测手段,来完全揭秘这颗不寻常的彗星(它不属于两类经典的彗星:木星家族和奥尔特云)(图片来源:ESA)

任务概念的一个主要方面是制造和操作大批量小卫星相关技术的挑战、合适的处理风险方法,以及定义一个吸引赞助方的有效框架(有可能是国际赞助方)。拥有国内支持和政府支持的私营企业正在为以低重复成本进行航天器大规模制造铺平道路,并提供了一种未来可效仿的商业模式。

可以设想,通过"泛星计划 2 望远镜"(Pan-STARRS2)与"泛星计划 1 望远镜"(Pan-STARRS1)协同合作①,以及借助近几年将建成的"大型综合巡天望远镜"(Large Synoptic Survey Telescope,LSST)等未来的望远镜设施,人类有能力在这些长周期天体到达近日点前至少 10 年发现它们。这个时间提前量完全足够我们研制和发射探测航天器,实现在长周期天体抵达其近日点的途中与它们相遇,而且这些天体大多需要穿越地球轨道。为了应对上文提到的挑战,任务设计框架应该

---

①　译者注:泛星计划(Pan-STARRS),全称为全景巡天望远镜和快速反应系统(Panoramic Survey Telescope and Rapid Response System),是一个引导对全天天体进行测量学和光度学基础研究的天文计划。2017 年 10 月 19 日,位于夏威夷毛伊岛上哈雷阿卡拉天文台(haleakala observatory)的泛星计划(Pan-STARRS)望远镜对奥陌陌星进行了观测。通过对比同一天区不同时刻的区别,该计划期望发现大量新的小行星、彗星、变星及其他天体。泛星计划的首个望远镜位于哈里亚基山顶(Haleakala,Maui Island)。该望远镜于 2008 年 12 月 6 日开始运行,由夏威夷大学(University of Hawaii)管理。

提前出炉。

## 2.2.2　发现系外行星

这一思路源于"弧秒空间望远镜天体物理学基础研究"（Arcsecond Space Telescope Enabling Research in Astrophysics，ASTERIA）卫星[1]，这颗6U立方星任务于2017年夏成功发射并由国际空间站释放。ASTERIA卫星基本上是一颗技术验证卫星，并为处于职业早期的科学家和工程师提供了培训机会。该任务引入了一些功能，包括长时间指向能力和测光监测能力，主要监测具有系外行星的宿主恒星。主要科学目标是寻找行星凌日现象，这种现象具体表现在恒星亮度上的变化。数小时的指向能力涉及大量其他天体物理应用，如恒星特性的测量。如果ASTERIA卫星技术得到充分验证，那么后续就可以发射数量众多的类似立方星的探测任务，每颗立方星将指向不同的恒星。这些立方星可以根据其测量需求不同而存在差异，如携带不同的滤波器（Cahoy，2015）。稍微大点的航天器上可能会使用更复杂的技术，如红外线或紫外线光谱仪。例如，最近选定的"恒星-行星活动研究立方星"（Star-Planet Activity Research CubeSat，SPARC）任务正计划通过紫外光度监测来评估恒星辐射环境（Shkolnik et al.，2018）。这类卫星任务概念的关键是一颗立方星只专注于一颗恒星。因此，如果在卫星运行上达成一致，那么国际合作就相对容易了，每个任务参与方都可以在他们认为合适的时候发射，并参加不同的任务阶段。此外，这可能也会允许民众科学地参与。

## 2.2.3　巨行星磁层和大气探测

利用立方星进行行星磁层和大气探测的想法是基于这样的预期：探测巨行星的大型任务可能会有发射余量，可以携带几颗立方星，并释放到这些巨行星的大气层或磁层中。NASA[2]已经将天王星和海王星这两颗冰巨星列为未来十年的主要研究目标。对这些行星的内禀磁场和磁层的认知是未来任务的重要科学目标。与地球相似，巨行星的磁层特征最好通过多点测量来实现。初步分析表明，与目前的此类测量方法相比，覆盖大范围经纬度并持续至少一个完整的自转周期（约10小时量级）的行星磁场同步测量将产生开创性科学成果。这些立方星可以按照一定顺序释放，以延长时间采样。高质量的磁强计已经小到可以放进3U立方星内（示

---

[1]　https://www.nasa.gov/mission_pages/station/research/experiments/ 2513.html。

[2]　https://www.lpi.usra.edu/icegiants/mission_study/。

例见 INSPIRE① 任务），而这个 3U 立方星还可以容纳用来进行重力场测量的应答机。这种类型的地球物理测量最理想的实现方式是各个立方星通过通信网络（立方星之间或立方星与母星之间的网络）实现同步测量。通信网络还具有其他优势，例如，飞越不同半球的立方星可以通过无线电掩星（radio-occultation）实现行星大气探测。

针对行星大气的另外一种应用就是在不同地点释放大量立方星来实现对化学成分（如挥发性、同位素）及其异常的探测。此类探测不需要立方星之间的通信网络支持。

尽管人们对此类冰巨星和其他行星的探测抱有浓厚兴趣，但是将立方星级别的航天器搭载在大型任务上的路径还不明朗。为了实现搭载目标，一种折中的方案是使立方星的研制遵守造价昂贵的大型任务的设计规则；但是这存在多颗立方星搭载成本过于昂贵的风险。现行的风险控制和任务保证方法可能使得在航天局研制任务上搭载国外单位研制的立方星更加困难。

上述概念研究涉及利用小卫星执行行星探测、系外行星探测及巨行星磁层和大气层探测。但是在小卫星执行这 3 项或者其他深空探测任务之前，仍然有一些技术障碍需要解决。如 1.2.3 节所述，小卫星在通信、能源、推进和任务运行等方面都面临特殊的挑战，并且这些挑战比地球轨道的卫星任务更大。

---

**研究发现 2.2**　小卫星为发射频率不高的行星际任务提供了机会，例如可以用于着陆器或作为“牺牲性”的卫星，以及通过小卫星网络实现“一生一次”型天体探测任务。

---

## 2.3　小卫星合成孔径望远镜

在天体物理领域中，很多基本科学目标要求对宇宙中最微弱的天体进行观测，并需要生成具有足够高空间分辨率的恒星和行星图像来解析其盘状结构。因此，探测任务要具有较大收集面积和/或较大有效孔径。

例如，NASA 将于 2021 年发射耗资约 90 亿美元的詹姆斯·韦伯太空望远镜。这个最大的实用孔径望远镜可以折叠后装进最大的火箭整流罩中。如果所需孔径

---

① https://www.jpl.nasa.gov/cubesat/missions/inspire.php。

增加一倍,下一代望远镜就需要采取不同的设计方法了(Sweeting,2018)。带有片段镜面的小卫星可以在太空中组装(或自动组装)成更大的结构(图 2.4,参见彩图;Saunders et al.,2017),甚至可以自由编队共同工作。

图 2.4　可重构空间望远镜自动组装(AAReST)任务概念:直径为 10 cm 的圆形镜面安装在一组立方星上[图片来源:加州理工学院 AAReST 团队(http://www.pellegrino.caltech.edu/ aarest1/)]

在不久的将来,专用的小卫星星座任务可以形成具有较大收集面积和/或较大有效孔径的合成孔径望远镜。新一代望远镜将具备行星成像、分辨恒星系统、探测近地小行星并成像的能力,其成本也将大大降低。

几十年来,射电天文学家利用合成孔径技术实现了较高空间分辨率和较大收集面积。因此,我们能够很好地理解和掌握从分布式望远镜阵列观测结果重构图像的方法。"超大望远镜干涉仪"(Very Large Telescope Interferometer,VLTI)和"高角分辨率天文中心"(Center for High Angular Resolution Astronomy,CHARA)阵列等地面观测设备已经验证了综合单个望远镜可以实现可见光合成孔径系统技术,这些望远镜采用了光束定向镜系统、真空管和自动相位延迟控制技术。将地面验证的技术应用到太空中,虽然很有挑战,但并不是不可能。

这种思路并不是新提出的。"空间干涉测量任务"(Space Interferometry Mission,SIM)的研究始于 1998 年,尽管进行了广泛研究,但是最终由于技术和成本原因而被放弃。正在研制的"激光干涉空间天线"(Laser Interferometer Space

Antenna,LISA)探路者任务是 NASA 和 ESA 的联合引力波探测任务。LISA 对指向精度的要求远远超过光学合成孔径阵列(optical synthetic aperture array)。LISA 探路者任务已经发射并超额实现了设计要求。

从 SIM 和 LISA 探路者任务中获得的共同经验是,小型原子钟和望远镜之间的光通信和精密干涉定位技术可用于建造 200 kg 的 1 m 孔径分布式望远镜阵列。使用脉冲光链路可以实现亚纳秒时钟同步(Anderson et al.,2018)。收集面积取决于阵列中望远镜的数量;一个有效直径为 10 m 的望远镜需要大约 100 个装有 1 m 望远镜的航天器,一个 30 m 的望远镜需要大约 1000 颗卫星。工业界为地球观测星座任务开发的新制造技术可以应用于此。如果大批量制造,每颗小卫星比较合理的目标成本价格约为 50 万美元。因此,一个 10 m 和 30 m 的分布式阵列(distributed array)的成本分别约为 5000 万美元和 5 亿美元。卫星设计和开发计划成本约为 1 亿美元,而卫星的发射成本和制造成本基本相当,因此,可以合理地预估 10 m 望远镜和 30 m 望远镜的总成本分别约为 2 亿美元和 11 亿美元。即使它们最终的总成本是预估结果的 3 倍以上,也会比"詹姆斯·韦伯空间望远镜"的成本低得多。

光子技术的发展为相控阵列光学望远镜(phased array optical telescope)的实现提供了另一种方法。随着新制造技术的发展,现在已经能够制造 1.2 m 平面相控阵列。第一个任务有可能是在一颗小卫星上搭载 1.2 m 的系统,而下一步可能是研制一个可折叠的 1 m 平面相控阵列。展开一组 9 个面板阵列(3×3 平面望远镜),收集面积为 13 m²,相当于一个 4 m 的望远镜,中间可视空间分辨率为 0.028″(角秒)。阵列面可以在每个光子波导上使用小透镜。而在每个透镜的表面都有一个纳米光栅用于产生随角度变化的谱移。望远镜是电控指向,因此可以得到波段的光谱图像。另外,为了获得高光谱分辨率,每个波导可安装一个可调的光子法布里-佩罗干涉仪(tunable photonics Fabry-Pérot interferometer)。阵列单元可以通过单模光纤连接到由计算机控制的光子相位延迟器上,此相位延迟器是航天器中相关器的一部分。

另一种有意思的可能性是,在航天器上安装一个 1 m 的光子望远镜,另一个安装在航天器可展开的长臂上(如 100 m 长臂)。尽管光学波段对航天器长臂的刚性要求不严格,但从结构和热稳定性的角度来看,这种设计确实是一个挑战。长臂的主要作用是提供和保持望远镜的相对位置。当长臂达到一定长度后,其面临的结构和热稳定性的挑战将大于自由编队飞行小卫星所面临的挑战,但是就现在来看,这个临界值还不得而知。

一个 100 m 的望远镜系统的空间分辨率大约是中心可视 $5×10^{-9}$ rad(弧度)

或者 1 毫角秒（milliarcsecond）。举例来说，在 1000 km 的近地轨道上，它可以分辨出地球表面 5 mm 的特征、月球表面 2 m 的特征以及 100 个地月距离一颗小行星上 19 m 的特征。它也可以分辨距离 10 光年远恒星上 $5×10^5$ km 的特征，以及太阳系附近恒星上的黑子，并可以用日震学技术来确定恒星内部温度、密度和自转频率等。

空间科学界非常希望"詹姆斯·韦伯空间望远镜"不负众望，能够从 2021 年开始产出突破性观测结果。但是我们很清楚，在后"詹姆斯·韦伯空间望远镜"时代，下一步的突破需要通过分布式系统来实现，但目前在技术上分布式系统似乎遥不可及。这不禁让人联想到欧洲南方天文台"超大望远镜"（VLT）的处境。它由 4 个 8 m 的望远镜组成，第一次观测是在 1998 年，但直到最近才有可能将望远镜信号合成一个 16 m 的望远镜信号[①]。在太空中，使干涉测量成为可能的第一步是小型射电望远镜，然后从红外波段发展到光学波段。与此同时，LISA 探路者任务在姿态控制方面取得进展，以及即将发射的 LISA 任务的足够高的姿态精度，能够实现对光学信号进行综合，因此最终可能会实现对另一个恒星系统中一颗类地行星进行成像。

> **研究发现 2.3** 在"詹姆斯·韦伯空间望远镜"之后，单孔径大型空间望远镜无法再继续发展。为了取得进一步进展，需要一种类似于分布式孔径小型望远镜的新方法。

# 2.4 星 际 任 务

目前，巨大的星际空间距离和人类有限的生命是阻碍星际任务的因素。要使未来的星际任务成为可能，就必须提高航天器的速度。而速度的提高，要么提高初始加速度，要么增加加速度的作用时间。

太阳帆（solar sails）的概念是利用太阳辐射压力作为卫星的推进动力。这样，卫星不需要携带推进剂箱，降低了系统质量和系统复杂度。但是，由于太阳辐射压力很低，因此必须具有较大的帆板面积和较长的加速时间。JAXA 成功研制的 IKAROS 金星探测任务于 2010 年发射，并验证了太阳帆技术（Tsuda et al.，2011）。"行星协会"（Planetary Society）的"光帆 1 号"（LightSail-1）任务于 2015

---

① https://www.eso.org/public/news/eso1806/。

年在轨道释放了一个太阳帆[1]。作为 QB50 项目卫星之一,于 2017 年发射的 3U 立方星释放了"充气帆"(Viquerat et al.,2015)。

激光帆(laser sail)与太阳帆类似,其推进光子是在地球上产生的。"突破摄星计划"(Breakthrough Starshot Initiative)是一项激光帆任务概念,由斯蒂芬·霍金(Stephen Hawking)、马克·扎克伯格(Mark Zuckerberg)和尤里·米尔纳(Yuri Milner)于 2016 年提出[2]。该计划旨在为下一代人类探测半人马座阿尔法星(Alpha Centauri)的任务奠定基础。这一任务概念以两个主要想法为基础:将目前航天器的总质量压缩到约 1 g;推进系统基于地面激光器。采用这样一个系统,使航天器速度达到光速的 20%是可行的(Lubin,2016)。凭借这么快的速度,航天器就有可能在空间研究人员的寿命周期内飞离太阳系开展行星际任务。

摄星计划所面临的技术挑战十分艰巨。计划的发起人也承认"要实现这些任务,还有很多艰巨的工程技术上的挑战等待解决",并列举了其中的 29 项;但他们同时表示"目前还没有发现终止项"[3]。例如,具备任务要求性能的光帆可行性还远远不能确定,因为这需要管理多个相互冲突的优先级要求,也需要一个部分满足所有优先级的技术解决方案(Atwater et al.,2018)。此外,纳卫星还将面临潜在的与星际气体和尘埃破坏性撞击的风险(Hoang et al.,2017)。与地球通信链路增益预算也很低,不论是光帆与地球直接通讯还是在链路上用卫星作通信中继,链路增益预算问题都需要解决。最后,这些任务会获得怎样的有用的科学测量数据还是一个未知数。尽管有很多挑战,前往另外一颗恒星的星际任务仍然是人类未来的终极愿景,值得进一步探索。

## 2.4.1 挑战和影响

将推进系统留在地球上的基本想法开启了新的一类任务和研究项目,但它们并不局限于星际任务。基于地球的激光推进基础设施也将使星际任务变得更快。同时,激光推进设施可能会成为前往未知天体任务快速反应系统独特的组成部分。

在星际任务发射之前,必须开发出一系列必要的技术,其中包括能够储存亿瓦时(GWh)能量的新能源系统,该系统与地面激光系统的通讯脉冲峰值功率高达 500 GW。其他还需要开发的技术包括高功率激光器、带有超反射涂层的超薄光

---

[1] http://www.planetary.org/explore/projects/lightsail-solar-sailing/。

[2] http://breakthroughinitiatives.org/initiative/3。

[3] 译者注:指该任务从科学原理的角度看不到任何致命障碍。

帆、更节能的通信系统，以及针对纳卫星的新集成技术。这些技术很多可以应用于陆地，从而为总体提升社会生活水平服务。例如，星际任务所必需的激光电源的研发可能会提高陆地发电厂和配电系统的效率，或者有助于解决可再生能源面临的能源储存问题。

## 2.4.2　星际任务实施之前

在星际任务实施之前，仅仅对航天器系统的测试和太阳系内飞行相关的设计就足以形成一个新的科学研究分支，如：

• 探测未知事件和天体的快速反应系统，如太阳系天体或者行星际小行星A/2017 U1"奥陌陌"的突然来临（Gaidos et al.，2017）。

• 使用集群卫星对小行星带天体进行 3D 测绘。

• 对日鞘层和终端激波的多点探测研究。

随着航天器速度接近几分之一光速，太阳系内的飞行时间将大幅下降。此外，一旦完成必要的地面基础设施的开发和建设，发射成本也就相当于地面基础设施的维护成本加上单个航天器的加速所需能源成本，因此发射成本也将下降。正如立方星所起到的作用，这可能为对空间科学感兴趣的更多人群打开了深空的大门。

地面激光推进系统由一个激光阵列组成，因此是完全可扩展的。这意味着若要增加加速能力，只需在激光阵列中额外增加激光器即可。这种结构也允许对任务场景进行反复试错。图 2.5（参见彩图）所示为同一激光阵列加速时，航天器的能量消耗、能量成本（0.12 € /kWh）[①]和最终速度随航天器质量变化的曲线。图中数据是假设了一个阵列面积大约是 $120 \times 120$ m² 、功率为 10 MW 的激光阵列，其航天器质量从 1 g～1 kg。航天器越重，加速就越慢，因此它们在激光器附近停留的时间就越长，从而增加了整体能量消耗和成本。激光阵列的大小是根据卢宾（Lubin，2016）给出的数值外推而得到的。

如果要追赶以每秒数十千米运动的天体，小型激光推进航天器是唯一的选择，并且造价足够便宜，能够在地球轨道上随时待命。尽管这一速度远远低于"摄星计划"的 0.2 倍光速（60000 km/s）的最终设计目标，但仍然可以作为不错的中间步骤。例如，如果图 2.5 中 1 g 的航天器被加速到大约极限速度的 1/1000，即大约

---

① http://ec. europa. eu/eurostat/statistics-explained/index. php? title＝File：Electricity_prices_for_non-household_consumers，first_half_2017_(EUR_per_kWh). png。

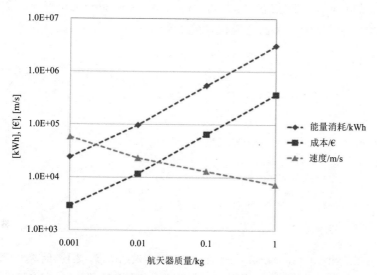

**图 2.5　航天器质量与激光阵列能量消耗、能量成本和最终速度之间的关系图**

60 km/s,那么它可能会在 8 天内与 A/2017 U1"奥陌陌"天体在最近的距离实现交会。这就需要及时发现"奥陌陌"天体(而事实上并非如此)。假设航天器发现"奥陌陌"天体后,马上穷追不舍,那么要追上它大概需要一个月的时间。图 2.6(参见彩图)所示为"奥陌陌"天体在抵达最近点之前的 1 个天文单位到最近点之后的 12 个天文单位之间的轨道。利用图 2.5 中质量分别为 0.001 kg、0.01 kg 和 0.1 kg 的 3 个航天器的最终速度,就可以计算出它们每天飞行的距离。在这些航天器发射升空之前(假定是 10 月 20 日,即发现"奥陌陌"天体后的第 2 天),航天器到地球的距离设定为 0 天文单位。通过计算"奥陌陌"天体与发射场之间的距离,我们就有可能估算出航天器追上"奥陌陌"天体的时间,如图 2.7(参见彩图)所示。从图上可以看出,10 MW 的激光阵列不具备将 0.01 kg 和 0.1 kg 的航天器加速到追上"奥陌陌"天体的速度的能力。若想解决这一问题,要么需要一个更强大的激光阵列,要么需要能更早预警的巡天系统(sky survey system)。按照目前欧洲的电价计算,加速每克航天器耗资约 3000 欧元。

在加速阶段之后,"芯片卫星"(ChipSats)进入滑行飞行阶段,这意味着任何轨道扰动(orbit perturbations)必定都是外力的结果。通过简单地追踪这些芯片卫星,就可以研究外力的性质。例如,在航天器上安装一个磁强计,就可以在保持数据返回可扩展的同时,研究其通过的空间的性质。成像仪器虽然可以进一步提升航天器的能力,但同时也提高了对数据带宽的需求。在卢宾的项目建议书(Lubin,2016)中,激光阵列既用于加速,也用于通信(采用光学阵列作为接收器)。另外一种可能是设想以其他深空探测航天器作为中继网络,就像 NASA 使用火星轨道器与火星车之间建立通信中继系统一样。与"洞察者号"(Insight)火星任务一同发射的两颗"火星立方星"(MarCO)已经验证了此中继系统。

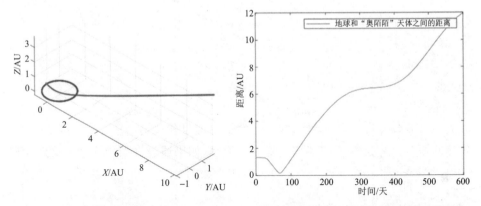

图2.6 左图:"奥陌陌"('Oumuamua)天体经过地球轨道时的轨道。右图:地球和"奥陌陌"天体之间的总体距离(本图所用位置数据取自 JPL Horizons 数据库和网页:https://ssd.jpl.nasa.gov/horizons.cgi。"奥陌陌"天体是达到最近距离后才被发现的)

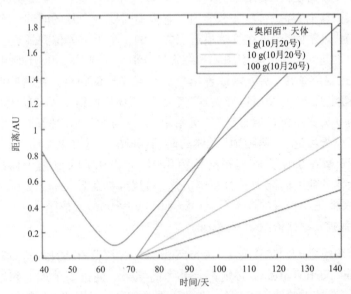

图2.7 蓝线代表了 2017 年 10 月 20 日"奥陌陌"天体与地球位置之间的距离。本图所示的模拟仿真也是从这个日期航天器所处的太空位置开始加速。红色、黄色和紫色三条线(参见彩图)分别代表三种航天器质量,都使用 10 MW 激光阵列进行加速。可见只有质量为 1 g 的航天器才能获得能够追上"奥陌陌"天体的速度

## 2.4.3　政治

星际任务的推进系统基础设施建设需要大量投资。根据输出功率的不同(1~100 GW),激光阵列的预计面积将在 $10^6 \sim 10^8$ $m^2$。该面积接近欧洲核子研究中心(CERN)的"大型强子对撞机"(LHC,直径 8.6 km),而且比欧洲南方天文台(ESO)的"极大望远镜"[①]要大得多,这两个项目都是国际合作和国际资金支持的成果。尽管还没有预估造价,但是为了获得项目所需的资金,很可能需要多家航天机构共同参与建设。"突破摄星计划"的最初项目建议书给出的建议是,根据任务需求选择一处地址安装大型激光阵列。但是,如果通过将大型阵列分成若干较小的阵列,并将国家政治考虑和任务需求相结合进行选址,则会使筹集项目资金变得更容易一些。和国际空间站起到的作用一样,将大型激光阵列部署在多个国家也必然会加强国际空间合作。但是,分布式激光阵列可能会带来激光相位处理问题。

激光阵列工作时将发出兆瓦的激光功率,从而危及任何在激光束内的物体,包括地球轨道卫星。根据任务类型和航天器大小,激光阵列的操作时间可能在 10 分钟到几个小时之间。鉴于在地球周围的轨道上有 4 万多颗卫星和其他物体,因此密切的协调对确保安全操作至关重要。与可能导致的潜在危险相反,激光阵列系统的推力也可以服务于许多具有国际意义的有利用途,如避免地球近地轨道航天器的碰撞(Stupl et al.,2012)和使小行星偏转(Thiry,Vasile,2014)等。

## 2.4.4　技术

互补金属氧化物半导体(Complementary Metal Oxide Semiconductor,CMOS)技术和立方体卫星展现的空间技术的发展表明,在保持一个系统的大部分甚至全部能力的同时,缩小该系统的物理尺寸和质量是可能的。虽然"摄星计划"是基于现有的技术,但这些技术中有很多还未应用到深空探测任务中,或者现在的技术成熟度还比较低。截至目前,卫星质量最低的技术验证卫星是基于单块打印电路板的 Sprite 卫星。Sprite 卫星由两次发射任务带入太空,但是还没有单独运行,即没有与母航天器分离[②]。因此,要促进"摄星计划"发展,这些技术就需要转化或者进一步研发。

---

①　https://www.eso.org/sci/facilities/eelt/。

②　https://www.scientificamerican.com/article/reaching-for-the-starsbreakthrough-sends-smallest-ever-satellites-into-orbit/。

未来激光波长设计为 1056 nm,即在红外光谱范围内。尽管在这个波长上地球大气层基本是透明的,但是山区一类的高地仍然是理想的选址位置。卢宾提出在完全开发的系统中,最终激光阵列的光功率在 1~100 GW。尽管也有激光器插座效率(wall plug efficiency, WPE)高达 50% 的报告(Pietrzak et al.,2015),但典型的激光器插座效率约为 20%(Botez et al.,2015)。因此,对最大的单一激光阵列方案来说,需要在附近配备 10 分钟内能供电 500 GW 的发电厂。相比之下,大型核电站的输出功率通常为 2 GW①,而航天飞机在起飞时的输出功率为 45 GW。显然,向这些激光阵列系统输入电力是基础设施方面的主要挑战之一,这将需要激光阵列所在国家或地区在政治上提供大力支持。对巨大的、几乎是脉冲式的能量的需求可通过把阵列划分成较小的子阵列来得到一定程度的缓解,不过这将带来一个新的挑战:对激光光束进行相位控制。

## 2.4.5　预计任务发射时间—— 一场技术赛跑

对微型星际航天器的需求,将带来或者加速航天器减重的趋势。这种趋势很可能是由于立方星的出现而兴起的。这里,将航天器版本的摩尔定律和牛顿第二定律相结合来预测这种趋势的速率,飞行时间 $t_C = (dm_{S/C})/(t_a F)$,其中 $m_{S/C}$ 是航天器的质量,$d$ 是航天器与目标恒星的距离,$F$ 是加速推力,$t_a$ 是推力 $F$ 的加速时间。

根据所知,目前还没有针对航天器减重速率的深入研究。减重速率是在航天器某种特定性能保持不变的情况下,航天器质量降低一半所需的时间。两份小型调查报告已经出炉,其中一份调查了 115 次卫星任务,共涉及 4 种任务类型(Fléron,2017),另一份调查了 47 项通信卫星任务(Fléron,2018)。第一份报告表明,地球观测任务的减重速率为 36 个月,但在 3 个深空任务上没有体现出明显的趋势。关于通信卫星的第二份报告给出的减重速率为 18 个月,这个数据看似很高,但是由于数据样本相对较小,所以可能并不准确。另外一种估算减重速率的方法是调查最简单的卫星——信标卫星。因为它们只有一个功能,所以容易进行比较。"斯普特尼克"(Sputnik)卫星、Cute-1 卫星和 Sprite 卫星都是信标卫星(图 2.8,参见彩图)。

根据上文,假设立方星已为航天器引入了摩尔定律,那么减重速率可以用来计算特定任务的最佳发射时间。为了说明这个过程,假设减重速率为 4 年,2000 年

---

① https://en.wikipedia.org/wiki/List_of_largest_power_stations#Nuclear。

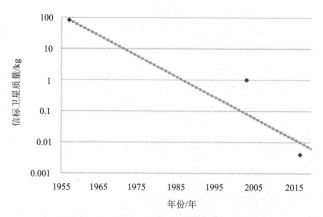

**图 2.8　Sputnik(1957)、Cute-1(2003)和 Sprite(2017)等信标卫星的质量演化。灰色趋势线表示质量减半周期为 55 个月**

初始质量为 1000 kg。所有的卫星任务都采用卢宾(Lubin,2016)所定义的地基激光阵列作推进系统。因此,加速推力 $F$ 在这几年中保持不变。第一个航天器的质量 1000 kg 是根据 2006 年发射的 478 kg 的"新视野号"(New Horizons)探测器和 2011 年发射的 3625 kg 的"朱诺号"(Juno)探测器而假设的。它的质量也大致与 20 世纪 70 年代发射的"旅行者号"(Voyager)探测器相当。目前,"旅行者号"已成为第一个星际探测器,但是要到达半人马座阿尔法星系统还需要 7.5 万年。根据飞行时间的公式,芯片卫星和类似的技术将第一个到达星际空间,即使大型航天器的星际任务不久就会发射,它们也会被芯片卫星半路超过,这是因为芯片卫星具有更高的滑行速度,如图 2.9(参见彩图)所示。

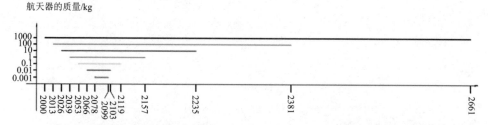

**图 2.9　航天器质量与发射、到达时间(目的地是半人马座 α 比邻星)的函数关系。假设航天器质量的演化遵循航天器版本的摩尔定律,这里的航天器质量减半时间被设定为 4 年(Fléron,2017)[航天器速度数据来源于卢宾(Lubin,2016)]**

　　上段最后一句话不应被误解为只建议在芯片卫星和激光发射系统足够成熟后再尝试星际任务。实际上,第一项星际任务是 1977 年就已经发射的"旅行者 1号",虽然当时并不是这么宣传的。目前,一项星际探测任务正处在论证阶段。虽

然它不是一颗小卫星,但它应该能在 50 年内到达 1000 个天文单位的深空
(McNutt et al. ,2018),其速度大约是"旅行者号"的 6 倍。国际空间委员会星际研
究小组(Panel on Interstellar Research,PIR)正在进行科学需求和小型设备技术研
究,指导此类星际任务在方式和实施上达成国际共识。尽管从目前来看,要到达另
一颗恒星的终极目标似乎还遥不可及,但任何尝试都将为未来的星际任务奠定基
础,并促进技术进步和科学发现。

> **研究与发现 2.4** 参与像摄星计划这样的愿景目标任务是令人激动的,尽
> 管该方案可能最后被证明不可行,但可以指导我们进行新尝试、寻找新想法和
> 发展新技术,而这本身就能产生很多应用。

# 第 3 章

# 小卫星未来发展的障碍和对策

　　本章将讨论从第 1 章"小卫星发展现状"到第 2 章"未来愿景"之间的发展障碍,并尝试找出克服这些障碍的方法和途径。本章将聚焦与科学技术障碍所对应的"体制障碍":科学界、航天机构、工业界和决策者(政府和国际组织)之间应如何合作,才能使空间科学小卫星带来的回报最大化?

　　高昂的任务成本(特别是发射成本,将在下文中详细讨论)也许是利用小卫星促进科学进步的最大障碍。它限制了科学卫星任务的数量,并导致各国空间机构在面对大型、昂贵的任务需求时,采取规避风险的态度。反过来,这可能会扼杀科学家的创新精神,因为任务评审过程和有限的任务建议机会可能会导致他们倾向于提交规避风险的任务建议书。这些问题可以通过商业航天的发展以及可分担成本和风险的国际合作来缓解。虽然国际合作还有其他困难(包括当任务涉及多个机构时任务决策的时间问题、法律背景的差异问题等),但核心挑战仍然是政府、科学界和工业界这三个领域之间缺乏协调。

　　政府机构在支持小卫星的发展以及促进创新方面发挥着关键作用。此外,政府机构有能力倡导科学友好的政策,通过宣传和制定法规来解决各种障碍,从而更科学地利用小卫星。它们也代表了领导和参与多国合作项目的主要机制。增加对国际合作的支持可以鼓励新参与者的加入,政府机构还可以作为技术知识中心提供技术服务,将任务开发方、验证与示范方、机构用户联系起来。

　　一些领域的进步对空间科学发展很重要,而工业界对推动这些领域的发展起到了重要作用。例如增加低价格、高可靠性零部件的供应可以降低空间科学小卫星的成本。小卫星大规模制造和测试方面的工业经验教训有助于使科学卫星星座任务变得更加可行。低成本进入太空的新模式（如小型运载火箭的研制和发射代理）也可能降低科学任务的发射成本。

最后，空间科学卫星充分发挥潜力还面临一些文化障碍。政府和科学界的文化对小卫星并不完全有利，他们的任务开发和管理方法也倾向于强调低风险和高可靠性。允许试验、冒险和失败的文化才能让小卫星事业蓬勃发展。

本章分为 5 节。首先，讨论政府机构在支持空间科学小卫星发展中的作用，包括如何解决资金和相关政策问题等；然后，讨论科学界和政府应如何利用工业界的发展，以及充分发掘小卫星的科学潜力而需要实现的文化变革；最后，讨论为了进一步开发空间科学小卫星以及为未来创新培养强大的人员队伍所需的合作模式。

# 3.1 资　　金

政府机构是为重要的科学任务和相关技术研发活动提供资金的主要资助方。NASA 和 NSF 在利用小卫星方面发挥主导作用，其中 NSF 特别关注立方星。这两个机构都是由科学兴趣驱动的。值得注意的是，早在 2008 年，NSF 就率先大胆地支持采用立方星标准进行科学研究。NSF 为"立方星计划"（CubeSat initiative）提供适度的资助（每个 3 年期任务约 90 万美元）[1]，在关键时期支持了一个新生的科学社区。也可以说，这种冒险精神和部署全新科学计划的意愿为空间科学立方星的革新铺平了道路。如今，NASA 的科学任务部（Science Mission Directorate）为科学立方星提供的资金已大幅增加，并且在其征集"探索者"和"机遇"（Opportunity）系列任务建议书时，都接受了立方星任务概念。

美国的其他机构，如美国国家海洋和大气管理局（NOAA）和隶属于国防部的各种组织，正在（或探索）使用非常小的卫星来完成任务。值得注意的是，尽管 NOAA 尚未部署政府拥有的纳卫星，但它正在向商业机构授予合同（目前正处于合同授予的第二阶段），以购买商业纳卫星星座收集的气候、大气和陆地成像数据[2]。

---

① https://sites. nationalacademies. org/cs/groups/ssbsite/documents/webpage/ssb_166650. pdf，检索日期 2019 年 2 月 23 日。

② https://www. nesdis. noaa. gov/content/noaa-continues-push-towardinnovation-partnerships-second-round-commercial-weather-data，检索日期 2019 年 2 月 8 日。

在欧洲,小卫星自 20 世纪 90 年代以来一直受到 ESA 关注。但是如 1.1.1 节所述,欧洲对于小卫星的资助非常零散。欧洲的个别国家和欧盟也做出了一些努力来解决这一问题,特别是欧盟"第 7 框架计划"(FP7)和"地平线 2020 计划"(Horizon 2020)提供了一些资助项目;瑞士也资助制造了"瑞士立方星"(SwissCube)。此外,一些国家的空间机构已经采取相关措施来支持国际伙伴关系的建立,例如以色列空间局(ISA)和法国国家空间研究中心(CNES)合作开展了地球观测和探索任务"植被和环境监测新型微卫星"(VENμs)。ISA 还与意大利空间局(ASI)合作开展了"星载高光谱陆地和海洋应用任务"(SHALOM)(见 1.2.2 节)。

在全球层面,防务分析研究所(Institute for Defense Analyses,IDA)发布的《全球趋势报告》(Lal et al.,2017)的附录中完整地列出了国际小卫星的活动和趋势。然而应该指出的是,迄今为止一些国家更多地关注小卫星的工业应用,而不是小卫星的科学计划(如 3.5.3 节将提到的"INDIA/ISRAEL@75"计划)。此外如 1.1 节所述和图 1.2 所示,大多数小卫星仅由少数几个国家发射。因此对多数国家来说,小卫星在科学上的应用还很不发达。

为了推动小卫星用于科学研究,政府机构除了资助完整的科学任务之外,还可以在其他方面发挥重要作用。例如,为开发跨领域技术建立特别机制,这样可以用更小的"任务包"实现更复杂的空间运载能力。根据拉尔及其他研究人员的研究(Lal et al.,2017),自 2013 年以来,NASA 的空间技术任务部(Space Technology Mission Directorate)已在小卫星计划上投资约 8000 万美元,主要用于开发小卫星星座、通信、机动和推进装置[①];值得注意的是,其中 60% 的资金流向了工业界。美国政府的其他机构(如国防部)也在技术开发方面进行了大量投资,这些投资也可能让科学任务获益。拉尔等人(Lal et al.,2017)还明确了政府机构应该投资的具体领域,包括综合研发领域,如卫星机动和推进装置、星座和自主运行、热控、通信、深空系统和航空电子设备、可展开系统、空间碎片减缓与控制技术以及科学界认为重要的其他技术;还包括可降低风险的投资(即提供在轨验证任务的机会),在可靠性测试和数据管理等方面"产业共享"(即对知识和能力的共享)的投资。在实现产业共享的过程中,政府机构应该尽可能地利用现有的组织机构,并借鉴成功模式。

各空间机构还能够落实更多的小卫星发射机会,这样一来概念验证型任务和专门的科学及应用开发型任务就可以获得新的机会,并且发射间隔可以变得更短(如从确立任务到发射仅需 3 年时间)。政府机构还可以鼓励相关任务概念的设

---

① https://www.nasa.gov/sites/default/files/atoms/files/nac_march2017_blal_ida_sstp_tagged.pdf,检索日期 2019 年 2 月 22 日。

计,以弥补卫星观测空白,确保关键空间测量数据的连续性(如对地观测等)。

各机构拥有推动技术实现标准化的共同愿望。但是如果要制定标准,各机构应认识到,标准化既可能有利于技术进步,也可能阻碍技术进步(根据具体情况而定)。因此应当鼓励各机构尽量制定宽松的标准,即只制定最必要部分,保留宽泛性和灵活性,以免阻碍创新。各机构还可以采取行动来促进自由和公开的竞争,而避免因为强加限制而造成对创新的扼杀,这样可以提高小型星载系统的能力。

除传统的国家级空间机构以外,新的筹资机制正在出现。高校以及高校成立的联盟(类似于那些为建造大型地基望远镜而成立的高校联盟)、私人基金会、私人捐助者和"Kick Start"项目等正在资助开发比立方星更大的空间任务①。然而,大部分资助小卫星的资金仍在美国,一部分在欧洲(主要是英国)。这很可能是因为其他多数国家正在将他们面向太空的资源集中于开发实用的通信和对地观测系统,而不是投资在科学任务上。尽管如此,鉴于小卫星的低成本特点,在没有大量投资的情况下启动空间科学计划仍是可行的。

> **研究发现3.1** 政府机构应该支持私营机构不太可能支持的特定领域。这些领域包括卫星机动和推进装置、星座和自主运行、热控、深空系统和航空电子设备以及空间碎片减缓和控制技术等。目前存在的一个关键问题是对共享资源和基础设施技术的支持不足,如数据库管理、可靠性测试设备和发射支持等。政府机构还应该更积极地参与需要整个科学共同体协调的活动,如标准化研究等。

> **研究发现3.2** 需要建立面向科学小卫星的资助机制,尤其是在美国和欧洲以外的地区。鉴于小卫星已被证明有助于科学进步,因此建立专门支持空间科学小卫星的资金流是十分有益的。

## 3.2 支持小卫星发展的政策的作用

除了通过技术开发和政府资助可以加强小卫星的科学应用之外,还有几个政策障碍需要解决,以确保更好地利用小卫星进行科学研究。在本节中,将讨论4个

---

① 如 ASU-Milo 项目、MethaneSat-Environmental Defense 基金和 BeyondGo 项目。

我们认为最关键的问题：频谱资源、出口管制、发射方式及与轨道碎片有关的限制。

## 3.2.1 频谱资源

用于向地球传输数据的无线电频谱以及可用于深空探测的特定频谱，对于任何空间活动都是至关重要的，并且是一种稀缺资源，至少目前如此（直到激光通信成为常态）。因此，频谱资源在国内和国际层面都得到了良好的协调和管理，任何空间物体（包括小卫星）未经授权发射任何类型的无线电信号都是非法的。国际电联（International Telecommunications Union，ITU）成员国批准的《无线电管理条例》（*Radio Regulations*）概述了如何使用无线电频谱，各成员国均在这一国际框架内管理本国无线电频谱的使用。例如，在美国，对于由 NASA 或其他科学机构拥有和运营的小卫星，国家电信和信息局（National Telecommunications and Information Administration，NTIA）通常授予其使用某个频率的权限，但不颁发许可证，该证由联邦通信委员会（Federal Communications Commission，FCC）进行颁发。在批准使用某个频率后，FCC 和 NTIA 将其频率分配情况提交给 FCC 的联络员，后者再提交给负责维护国际登记册的 ITU。科学卫星在所有拥有空间计划的国家都有专用频谱。然而，对于大多数科学立方星，包括 NSF 支持的立方星，到目前为止都是使用业余许可证和业余频段频率，但现在也变得越来越困难，在未来继续这样做也不现实。因此科学卫星现在必须获得实验用的 FCC 许可证，但目前还不清楚受资助的由高校拥有和运营的立方星可以使用哪些频率。

虽然在频谱问题上小卫星与较大卫星没有本质区别，但小卫星的开发和发射速度却超出了目前频谱分配与管理的协调能力。在美国，FCC 在授予许可证方面一直缺乏一致性和稳定性。有时 FCC 会批准一种样式的小卫星，但有时又不批准，即使这两种小卫星的设计是相同的[①]。对国际合作项目而言，因为需要协调多个国家的频谱系统，所以这一挑战更加严峻。

除此之外，因为许多商业运营商使用的频谱同时也是高校或联邦政府机构正在使用（或可能使用）的频谱，因此小卫星预期数量的大幅度增长[②]将对特高频、S

---

① 详情见：https://spectrum. ieee. org/tech-talk/aerospace/satellites/the-fccs-big-problem-with-small-satellites。

② 根据一些报告（如 Aerospace Corporation 公司的报告），在未来 10 年可能会有多达 20000 颗卫星被发射到近地轨道上，其中大多数质量低于 500 kg。不过也存在其他不同观点：根据 Northern Sky Research 公司的数据，在这个时间段内可能发射的卫星不到 4000 颗；Euroconsult 公司则认为将超过 6500 颗。

频段和 X 频段以及其他空间频段的分配协调工作造成越来越大的压力。

随着更多卫星的发射,频谱竞争将变得更加激烈:它不仅会发生在近地轨道卫星之间,也会发生在地球同步轨道卫星之间(例如,穿越赤道上空的近地轨道卫星将不得不改变波段,以避免干扰赤道同步轨道卫星,因为后者的频率权更优先)。为了实现频谱的共享使用,需要进行干扰分析和广泛协调等工作。作为工作流程的一部分,必须在频谱申请过程中完成频率协调,这对小卫星开发商来说是一个额外的风险因素。此外,随着"射频干扰"(RFI)的增加,现行的国家和国际上的无线电频率管理机制可能会进一步加强,这就要求科学界保持对不断变化的规定有所了解。另一方面,对陆基射电天文学来说,继续保护射电天文保护频段也是至关重要的,因此无线电频率管理对科学研究也有一定好处。

小卫星的运行管理者也必定面临与传统卫星运行管理者相同的其他问题,包括空间频谱使用与地基应用之间的竞争。目前考虑的供地基使用的几个频段(如作为 5G 通信中一部分 6 GHz 以上的频谱)与用于空间遥感的关键频段是相邻的。因此,卫星使用这些频段的能力可能降低,这也越来越令人担忧(Mistichelli,2016)。

还有一些问题是小卫星特有的。例如,获得频谱使用许可的程序漫长而复杂,这种程序在许多国家涉及多个机构之间的协调。大多数从事科学小卫星工作的研究人员通常不熟悉这些机构的角色和规则,有时在开发小卫星的过程中直到很晚才发现相关规定,由此导致可能被拒发许可证。小卫星开发商通常倾向于使用频率较低的谱段,因为相关设备更便宜,也更容易获得,但频率较低的谱段却是无线电频谱中最拥挤的部分。越来越多的小卫星需求可能会增加对更高频率带宽的需求,这就会带来一系列成本问题和挑战。监管机构也更愿意在备案时了解卫星轨道的细节;但对一些研究人员来说,这些参数要到开发过程的后期才能确定,如哪个合作方可以提供搭载发射小卫星的机会。

### 3.2.2　出口管制

大多数国家都制定了法律法规来保护他们认为与国家安全和外交政策目的相关的技术或产品,尤其是防止被他们视为对手的实体或国家获取。空间技术和产品中的关键部分,几乎总是具有军民双重用途。即使大多数空间技术和产品是为科学研究而设计和使用的,但原则上也可以用于军事目的。

虽然出口管制条例通常不适用于公共领域的一般科学、数学或工程活动(通常属于基础和应用研究),但由高校和其他科学研究机构人员进行解释往往很难。在一些国家,"视同出口"(指物项或信息提供给外国人)等概念往往难以理解和遵守,而且遵守这些法律的责任经常落在没有接受过此类培训的教职员工和学生的肩上。

受出口管制制度影响,政府和学术界之间正在就这些限制对合法科学活动的损害程度进行辩论。美国的高等教育机构辩称,过于强硬的出口管制法规可能会阻止最优秀的国际学生到美国学习,并阻碍项目国际合作。多年来,出口法律法规变得越来越复杂,政府机构的执法力度也越来越强。在美国已经有大学人员因违规而被起诉,相关信息是可以公开查到的。对科学家来说,他们需要在国际合作和确保他们的研究工作符合出口要求之间进行平衡,在某些情况下,这会阻碍科学家参与国际合作。

过度的出口管制法规是否一定符合一个国家的最佳利益也没有定论。根据立方星的发明者、斯坦福大学空间系统开发实验室的教授鲍勃·特威格斯(Bob Twiggs)的说法:"ITAR(或国际武器贸易条例,即美国出口管制制度)正在把科学研究赶出美国,孤立美国,并导致相关市场在美国之外的地区发展。也就是说,那些处于全球导航卫星系统、电子学、控制系统和火箭系统前沿的外国学生就不能在美国进行研究了。"[1]美国国家空间委员会(National Space Council)的执行秘书说,"(一些)限制和过时的参数可能会产生意想不到的影响,即激励航天产业向海外转移,或导致制造商改变其供应链,从而损害国家安全。"[2]同样的情况也适用于科学小卫星企业。

小卫星为国家间的合作提供了独特的机会,远比传统空间活动提供的机会更多。小卫星项目确实可避免许多出口管制条例的严格限制,因为它们使用了商用货架产品和其他的主流部件,并且几乎不使用敏感技术。但是,目前并没有正式的规定将空间科学小卫星排除在管制条例之外。

显然,通过防止转让关键技术,出口管制原则上在加强国家安全方面发挥了有益的作用。然而,如果它所包含的条款在防止知识转让方面过于强硬,可能会损害长期国家安全,因为其阻碍了知识的自由交流,而在日益全球化的科学共同体中,知识的自由交流对于空间研究取得成功至关重要。任何管控小卫星出口的制度都需要在促进科学进步(包括通过强有力的国际合作)与不损害国家军事和经济发展利益之间寻找平衡。这种平衡需要通过小卫星的空间研究人员和决策者之间定期

---

① https://www.satellitetoday.com/telecom/2008/08/01/itar-balancing-the-global-play-field/。

② https://www.hudson.org/research/14341-full-transcript-space-2-0-u-s-compitativility-and-policy-in-the-new space-era。

和稳定的对话来确定,包括定期修改小卫星界感兴趣的技术清单等(Broniatowski et al.,2005)。更好地澄清规则和条例,包括更明确的解释,将大大有助于确保人们尤其是空间科学小卫星从业者遵守法律及其背后的精神。

## 3.2.3　发射

在过去,小卫星主要通过以下三种方式发射:在发射主星(如运往国际空间站的卫星或货物飞船)的火箭上搭载发射;通过"集群发射"(cluster launch)方式与其他小卫星一起发射[如 2017 年印度"极轨卫星运载火箭"(PSLV)发射的 104 颗小卫星];购买专用小型运载火箭(如轨道 ATK 公司的飞马座火箭)发射。迄今为止,大多数小卫星发射都是采用搭载方式。

在美国,NASA 通过"立方星发射计划"(CubeSat Launch Initiative,CSLI)下的"纳卫星教育发射"(Educational Launch of Nanosatellites,ELaNa)项目来支持科学小卫星的发射,包括商业发射和其他发射途径,例如向国际空间站进行货物补给的商业发射任务,以及"空间发射系统"(Space Launch System,SLS)的"EM-1 飞行任务",后者搭载了用于科学和技术验证的 13 颗立方星。除了政府,科学家还有其他支持来源:"联合发射联盟"(United Launch Alliance,ULA)公司为高校立方星提供竞争性的免费搭载服务[1]。为了方便高校与发射供应商建立联系,代理商提供发射协调和打包服务,如 Spaceflight 公司和 TriSept 公司等。

目前,全球有 100 多家发射公司致力于开展小卫星发射业务[2]。虽然这些运载火箭中的大部分可能不会获得最后的成功(它们大多数还处于开发阶段),但这一领域显示出了过去在小卫星界或发射界所没有的活力。对科学用户来说,有如此多的可获得、低成本的途径进入不同空间轨道,是令人欣慰的。

尽管拥有这些机会,但仍有很多低成本发射科学小卫星的需求是被压制的。截至 2018 年 12 月 1 日,NASA 的"立方星发射计划"支持的 162 颗立方星中仅发射了 66 颗,另外 38 颗已经交付发射商[3]。这一计划下的"风险级发射服务"(Venture Class Launch Services,VCLS)项目正在推动利用专用运载火箭发射立方星来减少积压情况,例如使用荧火虫太空系统(FireFly Space Systems),以及火

---

①　http://www.ulalaunch.com/ula-likes-transformation-cubesatlaunch.aspx。

②　https://www.spaceintelreport.com/count-em-101-new-commercialsmallsat-dedicated-launche-vehicles-in-development/。

③　https://www.nasa.gov/content/cubesat-launch-initiative-selections。

箭实验室（Rocket Labs）公司和维珍银河（Virgin Galactic）公司的运载火箭。Rocket Labs 于 2018 年 12 月 16 日执行了 VCLS 项目的首次发射，共向近地轨道发射了 13 颗立方星（其中 10 颗属于 VCLS 项目）。

目前，为小卫星界提供服务的新发射选项正在出现（Cappelletti et al.，2018）。已有 50 多家公司正在开发小型火箭来发射小卫星（Sweeting，2018）。然而，火箭技术开发是众所周知的高风险事业，很多活动开发都有可能失败。此外，小型火箭的特定成本往往较高；虽然大型火箭价格昂贵（图 3.1，参见彩图），但作为搭载发射选择，它们提供了最高的性价比（图 3.2，参见彩图）。例如，6200 万美元成本的"猎鹰 9 号"火箭可发射载荷 22800 kg，即每千克 2720 美元，这比"飞马座 XL"火箭发射 1 kg 的成本低了近 50 倍。但是搭载发射能够提供的灵活性是最小的，如运行轨道和轨道倾角的选择，甚至包括搭载卫星是否具有推进能力等；而这些因素又是开展良好空间科学研究的重要决定因素。

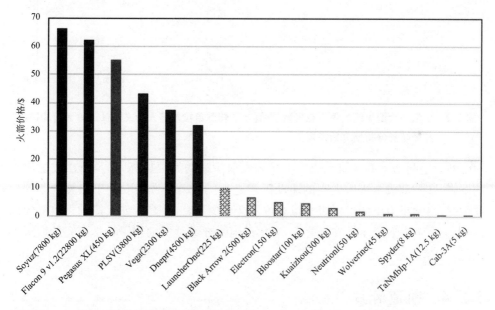

**图 3.1　火箭价格与最大发射能力对比（按成本排序）。黑色柱状图代表大型火箭，红色代表小型火箭，阴影代表研制中的火箭［本图转载自拉尔等人的文章（Lal et al.，2017），并经防务分析研究所许可］**

无论如何，发射仍然是限制小卫星发展的瓶颈。如果小卫星的数量和功能如预期一样增长，那么就需要加强低成本发射的机会。虽然已经有一些新的大型火箭亮相，如"蓝色起源"（Blue Origin）公司的 New Glenn 火箭、SpaceX 公司的"猎鹰重型"(Falcon Heavy)火箭、ULA 公司的"火神"(Vulcan)火箭等,这些大型火箭可以支持科学小卫星的发射,但首先需要解决小卫星的集成搭载问题。

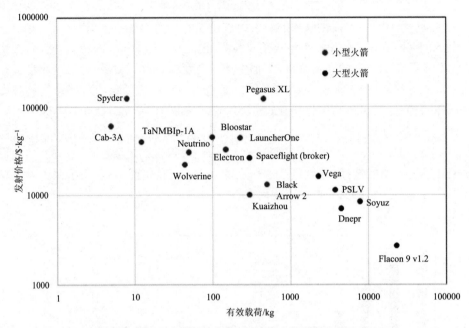

**图 3.2　小型和大型火箭的每千克发射价格[本图转载自拉尔等人的文章(Lal et al.,2017),并经防务分析研究所许可]**

商业空间活动的发展和成功将导致发射频率的大幅提高,以及单次发射成本的下降,这是低成本进入太空的驱动因素之一。但另一方面取决于相关产业的发展情况,科学小卫星的发射机会也有可能会减少。其他因素(如火箭的可重复使用性)也会影响发射成本。在此之前,各国政府需要继续资助用于科学研究的航天器发射。

## 3.2.4　轨道碎片

太空正变得越来越拥挤,预计近地轨道卫星数量的增长将由小型商业卫星(而非科学卫星)主导。预计 2017～2026 年将发射 3600～6200 颗甚至是 25000 颗小卫星(质量小于 500 kg 的卫星)①。令人担忧的不仅仅是卫星数量的增加,还有它们作为碎片留在太空的时间(取决于其轨道高度)将比它们的使用寿命更长。

随着航天器数量的增加,它们相互碰撞的概率也会增加,特别是当地面无法很好地跟踪,或者无法进行轨道机动时。由于近地轨道上物体的速度很高(约 10 km/s),所以即使是亚毫米级别的碎片也会对航天员和无人探测任务构成现实威胁。近地轨道的小卫星特别是立方星正日益被视为未来几十年的主要轨道碎片来源(Bastida Virgili,Krag,2015;Matney et al. ,2017)。尽管科学小卫星所占的比例很小,但它是这一挑战的一部分,而科学界必须面对这一情况。

与碰撞有关的担忧主要集中在计划建设的"巨型"通信星座上。航空航天(Aerospace)公司最近的一份报告评估了增加两个大型星座(分别来自 SpaceX 公司和 OneWeb 公司)对现有近地轨道星座(Iridium、Orbcomm 和 Globalstar 星座)的影响。该报告发现在新增的首个大型星座进入轨道的前 20 年内,预计每年会造成一次碰撞;在发射约 190 年后,碰撞率会达到峰值,可能增加到每年约 8 次。

随着小型航天器(特别是近地轨道上的小型航天器)数量的增加,对运营方的限制措施可能会越来越多,其中包括对科学卫星运营方的限制,即使科学航天器并不是即将到来的空间碎片挑战的根源。一些限制很可能是仅仅针对立方星而不是一般小卫星的,因为到目前为止,立方星通常不安装推进装置,在没有主动发射信号时无法被跟踪,也没有机动能力。这些限制措施尤其可能针对以下 3 个方面:① 确保所有小卫星都可以被主动或被动跟踪;② 减缓射频干扰(RFI);③ 在卫星停止运行后,遵守更严格的离轨要求。关于最后一点值得注意的是,近年来许多专家开始认为,建议立方星在其运行结束后 25 年内离轨的国际准则已经过时,可能需要更新。

科学界有机会通过积极寻求技术解决方案来避免未来可能出现的一些问题,如通过低成本手段使立方星具有可机动性、可跟踪性、避免射频干扰和及时离轨等特点。为了评估哪些方案具有成本效益,可能还需要更多的研究与开发活动。

## 3.2.5　总结与发现

为了有效地利用小卫星进行科学研究,需要应对 4 个关键的政策挑战。第一,

---

① 　如果 160 多个星座(其中大部分可以由小卫星构成)全部建成,那么近地轨道上将会有超过 25000 颗卫星(这是一种不太可能发生的情况)。这些卫星中的 90% 将主要用于通信,其余多用于对地观测和遥感。

频谱是一种稀缺资源,小卫星科学界不仅需要更好地了解航天器频谱分配的过程(可能很耗时),还需要了解该领域不断涌现的快速变化。第二,许多国家的出口管制法律会给科学家带来不必要的负担,遵守复杂的制度会抑制科学合作;在这方面同样也十分需要对科学界进行相关的培训。第三,发射成本是影响小卫星科学研究的关键抑制因素。政府机构通常会资助小卫星的发射,这种做法需要保持。但需要指出的是,低成本的空间飞行更需要大型火箭,由于专用于小型航天器的小型火箭的每千克成本高于搭载发射成本,因此小型火箭仅能服务于需要对轨道参数进行额外控制的细分市场。第四,随着空间航天器数量(尤其是近地轨道)的增加,对卫星运营方的限制可能会越来越多,包括对科学卫星运营方的限制。这些限制可能与卫星跟踪、频率干扰以及更严格的离轨机制和碎片减缓等有关,科学界需要积极主动地应对这些挑战。

> **研究发现 3.3** 频谱资源(用于向地球传输数据和科学研究)对于任何空间活动都至关重要,而且是一种稀缺资源。

> **研究发现 3.4** 科学研究的合作正受到与国际交流和合作相关的法律和规章的遏制,为了遵守这些法律和规章已造成了不必要的额外负担。

> **研究发现 3.5** 通过容易获得的搭载方式实现低成本发射,已经成为推动小卫星科学研究发展的关键因素。

> **研究发现 3.6** 随着空间航天器数量(尤其是近地轨道)的增加,对卫星运营方的限制可能会越来越多,包括对科学卫星运营方的限制。这些限制可能与卫星跟踪、轨道机动性和轨道碎片减缓等有关。

## 3.3 利用工业界发展

目前对科学研究的资助仍然主要来自政府机构,但科学界可以通过利用商业界的力量来降低成本或提高发射能力,特别是以小卫星为重点的新兴产业力量。例如,如果 NASA 近地轨道"小型探索者"(Small Explorer)任务的成本能降到2500 万美元的水平,即降低 10 倍,那么发射机会将大大增加,同时也会为科学研究创造新的可能性。不断增长的小卫星产业也有助于吸引和留住有才华的科学家和工程师,帮助建立空间科学和航空航天工程师队伍。

小卫星的研发由私营机构(包括大学)主导,即大多数小卫星是由私营机构或商业组织发射的。根据拉尔等人(Lal et al.,2017)统计的包含 650 多个小卫星机构的数据库,全球超过 50%和美国超过 75%的小卫星机构属于私营部门。虽然近年来,学术领域对小卫星的使用有所增加(图 3.3,参见彩图),但商业运营的小卫星仍占主导地位。仅在过去 6 年中,就有超过 475 颗商业小卫星发射(Halt et al.,2019);大多数商业小卫星被用于遥感测量(见图 1.3;Halt et al.,2019)。随着商用巨型星座的出现,预计更多的卫星将侧重于太空宽带服务①。

科学界有几种方式可以利用商用领域的发展。3.2.3 小节已经讨论了小卫星发射的方式。本节将讨论科学与工业合作的方式。

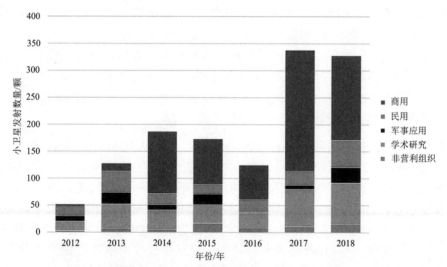

**图 3.3  各领域发射的小卫星数目[图片来源:布莱斯空间和技术公司 (https://brycetech.com/reports.html)]**

## 3.3.1  商用货架产品(COTS)零部件

在工业界,商业对地观测和通信都是小卫星利润最丰厚的"杀手级应用"(killer-apps)领域,也是当今大多数商业小卫星的应用领域。在这些领域不仅出现

---

① 例如,前往 https://www.nsr.com/smallsat-growth-on-shaky-foundations/查看根据应用分类的小卫星项目。

了越来越多的运营商(如 Spire 公司、Planet 公司等),也出现了越来越多的部件制造商/供应商(如 Gomspace 公司、ISIS 公司和 Blue Canyon 公司等)。他们以降低成本为目标,专注于大批量生产。对科学界来说,虽然较低的成本很重要,但更重要的是获得具有航天质量的商用货架产品零部件。随着大型卫星星座的出现,这一趋势正在加速(至少有 16 家公司专注于将卫星星座用于对地观测或天基互联网)。但是至少需要卫星平台的商品化,才能建成这些星座。许多公司都在借鉴非航天产业的方法和技术,例如采用视差算法(parallax algorithm)的小卫星群近距离操作,这种算法类似于为汽车防撞系统开发的算法。为了进一步降低成本,许多制造商和运营商正在试验采购商用货架产品零部件应用到自己的系统中。

## 3.3.2　商业数据采购

私人投资额度也可能超过政府投资额度的一个量级以上(至少与非涉密的政府投资额度相比)。因此,科学界应密切关注私营部门的发展,不仅要采购产品,还要采购服务。商用小卫星也可能收集与科学研究相关的数据,在对地观测和空间天气研究方面尤其如此。NASA 和 NOAA 已经开始小规模的试点计划,来观察工业界能否按照他们的标准生产数据产品。希望这将形成一个新局面,即政府可以购买数据,而不仅是建造和运行昂贵的卫星系统。

天气预报是目前正在发展中的一个应用领域。一些天气预测模型利用包括全球定位系统(GPS)的全球导航卫星系统(GNSS)的信号进行射频掩星测量,这些模型的价值已经得到验证。GPS 掩星计划得到中国台湾空间组织、美国 NOAA 和NASA 及私营企业的资助。政府希望通过购买数据来降低天气预报的成本,并提高预报的准确性,而这可以通过卫星星座实现,因为星座能够提供更密集的数据集和更短的重访周期。然而,商业开发者可能不满足于仅仅向政府提供数据服务。通过政府和私营企业之间的合作沟通,被政府购买和公开发布的数据可能也可以被私营企业用来生产裁剪后的产品,以满足特定客户的需求;特别是得益于大型计算机和新颖的软件系统,定制近实时的气象数据的价值正在增加。因此政府购买商业数据的想法能否实现,还取决于小卫星开发商发现哪个更加"有利可图":是为商业客户制造的产品,还是为政府机构制造的产品。

目前,自由和开放的数据政策对科学界来说非常有价值,因为它让政府机构产生的数据集在科学研究中有了新用途。开放数据政策也增加了对国际数据的利用。但应该注意的是,如果追求政府机构与工业界的数据购买伙伴关系,就有失去这种开放数据政策的风险。尽管商业数据带来了潜在的新机遇,但科学界和政府机构都必须努力确保开放数据政策不会在签订商业合同时受到影响。

### 3.3.3　代管有效载荷

"边缘与盘面全球尺度观测"（Global-scale Observations of the Limb and Disk, GOLD）任务是 NASA 第一个在商业卫星上作为"代管有效载荷"运行的科学仪器①。GOLD 任务于 2018 年 1 月 25 日搭载 SES-14 卫星升空，并于当年 6 月抵达地球同步轨道。它的测量结果（图 3.4，参见彩图）将提高我们对地球最外层大气层的认识，这对研究空间天气至关重要。该卫星能在同步轨道上运行，很大程度上是因为它利用了"代管有效载荷"这一途径②。GOLD 任务每 30 分钟扫描西半球一次，使我们能够首次监测高层大气的日常变化。越来越多的商用星座，特别是在近地轨道的商用星座，有可能为"代管有效载荷"提供新的机会。

### 3.3.4　工业界和大学的合作

工业界的参与是创建小卫星生态系统的重要一步。公司和高校合作推进知识的边界可以形成强大的创新引擎。对学术界而言，这种合作的优势是不言而喻的：它可以让学生和研究人员从事开创性的研究，拥有获得外部资金的更大潜力，在教学内容中引入更前沿的进展，为紧迫的全球挑战提供解决方案，等等。工业界也可以从劳动力的发展和培训以及招聘潜在员工等方面受益。

这种伙伴关系的一个最近的例子是"纳卫星应用和操作基准工程验证"（Nanosatellite Applications and Operations Bench for Engineering and Demonstration, NANOBED）项目，这是一个由英国思克莱德大学（University of Strathclyde）牵头、与 Clyde Space & Bright Ascension 公司合作开展的项目，目的是开发一种用于研究、创新和技术开发的工具。该任务支持快速的端到端的立方星任务设计和技术开发，技术成熟度最高可达到 6（TRL 6），其内容包括：① 定制任务和系统设计软件；② 集成硬件和测试；③ 通过模拟无线电通信线路实现硬件通信；④ 通过日常场景模拟运行；⑤ 在模拟运行场景中验证硬件功能。

利用任务和系统设计软件，工程师可以对以下任务内容进行定义：卫星轨道、地面控制站、航天器设计、工作模式和模式切换条件，以及来自文件库或用户定义

---

① http://gold. cs. ucf. edu/gold-will-revolutionize-our-understanding-ofspace-weather/，检索日期 2019 年 2 月 28 日。

② 译者注：因为同步轨道高度较大，发射至这一轨道成本很高。

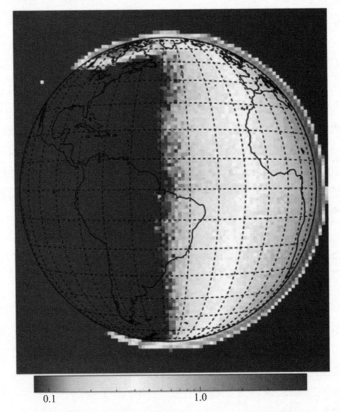

0.1                    1.0

图 3.4　NASA 的"边缘与盘面全球尺度观测"(GOLD)任务捕捉到的第一张地球高
　　　　层大气层紫外原子氧释放图像(135.6 nm 级)[图片来源:NASA/大气和空
　　　　间物理实验室 GOLD 任务科学团队]

功能的子系统,如"可扩展标记语言"(XML)等。在此基础上,工程师还可以对任务内容进行模拟,如轨道和姿态动力学及控制、地面跟踪可视化、太阳电池板的功率分布模式和相关的电池电压、地面站可见性、数据收集和数据下行等。

　　软件被安装在硬件平台上,这个平台应与飞行设备有接口,并且带有数据获取单元,能够获得用来检测电压的数据等。任务和系统设计软件的输出可以驱动电源包络设计,同时它反映了太阳能电池板的尺寸设计,并可提出硬件设计更改,包括当 GNU Radio 模块[①]与上行和下行链路信号链接时触发地面站工作。核心硬件设置如图 3.5(参见彩图)所示,其中额外的空插槽可以添加其他硬件,如有效载荷控制单元等。

———————————————

　　① 　译者注:GNU Radio 是免费的软件开发工具套件,可提供信号运行和处理模块。

**图 3.5**　连接飞行硬件的纳卫星应用和操作基准工程验证(NANOBED)项目的核心硬件装置,位于斯特拉斯克莱德(Strathclyde)大学实验室[图片来源:马尔科姆·麦克唐纳(Malcolm Macdonald)]

NANOBED 项目已被推广到世界各地的许多高校和研究机构,如墨西哥、美国和英国的一些高校和研究机构,还有一些将被部署到包括南非在内的其他国家。它创建了一个全球合作者网络,可以通过培训课程非正式或正式地协同工作,并相互分享经验教训。NANOBED 为学术界内部以及学术界和工业界之间的合作提供了一个平台。

最富有成效的合作形式应当是各参与方实现仅靠他们自己很难或无法做到的事情,这种合作形式可以围绕一个共同的研究愿景而建立,并且可以持续 10 年或更长时间,以此建立深厚的专业联系,相互信任,共享利益,从而弥合学术界和工业界之间的文化鸿沟。长期的合作还可以发展人力资源,而这是让学术界和工业界合作发挥作用所必需的。随着时间的推移,良好的合作伙伴关系将会培养出越来越多的教授和研究生,他们能够跨越文化鸿沟进行思考和行动,与公司的主要研究领域建立联系,制定共同的战略目标。

> **研究发现 3.7**　现成货架部件的采用为科学小卫星的建设和运行提供了一种完全不同的方式,并且增加了任务的弹性。采用大规模生产技术可以创造出快速、创新和低成本的新航天系统。

**研究发现 3.8**　越来越多的商业小卫星星座可能通过商业数据销售、代管有效载荷、搭载发射等方式为科学研究提供新的机会。然而,这也对当前的开放数据政策带来了风险。

**研究发现 3.9**　小卫星产业为那些对航空航天科学职业感兴趣的学生提供了有用的培训场所,尤其是工业界和学术界的伙伴关系,有助于确保未来有一支强大的航空航天和空间科学人才队伍。

# 3.4　支 持 创 新

小卫星领域最近的发展为科学研究提供了巨大的机会。为了最大限度地实现这一潜力,科学家和资助机构需要认识到小卫星不仅仅是大型卫星的微缩版本,它还可以提供新的测量方法,可以用"飞行—学习—再飞行"的模型来开发,并且成本较低。事实上,质量并不是小卫星的决定性特征,其决定性特征是它所产生的独特文化:与强调卓越能力、长寿命和高可靠性系统的传统机构相比,这种文化与鼓励冒险和快速创新(甚至以牺牲任务安全性为代价)的技术初创企业有更多的共同点。这类更小、更灵活的新机构所特有的新方法正在被工业界所接受,但是它们和大型空间机构①之间存在着文化差异,因为后者在很大程度上仍然采用传统的方法。要使小卫星驱动的科学走在前沿,还需要一种新的范式。

在《三箱解决方案》(*The Three Box Solution*)一书中,维贾伊·戈文达拉场(Vijay Govindarajan)(Govindarajan,2016)强调,需要采用不同的方法来解决任何机构面临的 3 个相互竞争的挑战:① 在当前保持卓越;② 识别并放弃过时的实践方法;③ 产生能够引领未来产品或方向的突破性想法。简而言之,为了成功地应对这些领域的挑战,机构需要将完全独立的资源投入到各个领域;创新需要不同的技能、标准、方法和管理策略。

政府机构虽然不同于初创企业,但适用同样的原则。美国国家研究委员会(National Research Council)2010 年的一份报告指出:"……DOD② 的经验表明,管

---

　① 还有像洛克希德·马丁(Lockheed Martin)公司这样的传统航天公司,它们的运作更像是一个航天机构。

　② 美国国防部。

理基础研究类型的活动所需的程序、衡量标准和管理技术,与管理先进技术开发和系统验证活动所需的通常是不同的。"同一份报告还指出,NASA 的大部分活动都集中在风险控制上,这对大型且昂贵的飞行任务来说是必要的;但另一方面,那些促进任务立项与创新研究的活动则需要不同的管理策略。

鉴于小卫星可能蓬勃发展所需的独特文化,效仿高风险研究(珍惜失败的价值)的组织模式可能是有用的。美国和其他国家的一些机构已经试图建立一种用于开展风险更高的科学研究(但科学回报也可能更高)的组织架构,它们可以作为在大型机构中培育小卫星(比传统的空间平台风险更高)的范例。

美国一些先进研究计划局(ARPA),如美国国防部先进研究计划局(DARPA)、美国情报先进研究计划局(IARPA)、美国能源部先进能源研究计划署(ARPA-E),已成功创建了这样的组织框架。在这个框架中,领导层优先考虑高风险的、不一定有明确定义的计划和项目,以维护这一架构的高风险文化(Peña et al.,2017)。① 同样,英国工程和物理科学研究委员会(Engineering and Physical Sciences Research Council)在名为"创意工厂"(IDEAS Factory)的框架中提出了各种方法,旨在刺激高度创新、高容忍风险的研究活动,这些活动在正常情况下是难以想象的。②

美国先进研究计划局使用一种特别的组织架构即"岛 + 桥"模型(Bennis,Biederman,1997;Sen,2014),抵制小卫星文化的大型机构可以借助这种模型来培育小卫星活动。在该模型中,"岛"是实验和失败的避难所,而"桥"是知识和技术向用户转移的渠道:新的技术能力需要发挥"出去",而外界的需求和其他类型的影响则必须"进来"。研究活动既不会像传统实验室那样完全脱离现实世界,也不会受制于现有的运营商。"岛 + 桥"模型将"互联科学"的方法应用于研究活动,将技术推动和需求拉动进行结合,对隔离和连接的需求加以平衡,从而为杂乱无章的过程提供合适的架构类型(Sen,2015)。

利用这个模型,可以在更大的空间机构中创建专门针对小卫星活动的子机构,以确保小卫星不被忽视,同时保证与更大的组织架构的联系(以确保发挥新的技术能力,满足需求)。该模型既不是新出现的,也不仅限于政府机构。在工业界,洛克希德·马丁公司的"臭鼬工厂"和 IBM 公司的"个人电脑"项目都成功使用了"岛 + 桥"模型。

---

① https://www.ida.org/idamedia/corporate/files/publications/stpipubs/2017/D-8481.ashx。

② https://epsrc.ukri.org/funding/applicationprocess/routes/network/ideas/experience/。

**研究发现 3.10** 让小卫星蓬勃发展的文化是一种允许试验、冒险和失败的文化。传统的空间机构倾向于强调低风险和高可靠性的空间系统,这种机构要培育小卫星活动就需要进行艰难的文化变革。为了确保实现这种变革,工业界甚至政府机构都有成功的模型可以效仿。

# 3.5 合 作

项目越大、越复杂,就越需要共同合作研究与开发来实现项目目标。例如,欧盟 QB50 项目,以及"国际远程信息处理任务"(Telematics International Mission, TIM)都属于具有复杂架构、运用多种测量手段的卫星星座。尽管 QB50 联盟不太可能在计划资助结束后继续存在,但参与该联合研究计划的高校在未来更有可能会继续合作。正在增长的多国科学合作活动也将支持新参与者的加入,并且将他们与任务开发者、验证任务、机构用户联系起来。这种合作在共享资源与分担风险的同时,也增加了对科学做出重大贡献的机会。这也有助于确保执行研究任务的是那些具有最适当经验(或拥有兴趣和需求)的科学人员。

## 3.5.1 合作模式

### 3.5.1.1 TIM 任务案例研究

在 TIM 案例中,各合作伙伴分别提供单独的卫星,共同组成一个星座,以便从该星座所产生的更大数据库中获益。7 个"区域领导人峰会"(RLS)[①]的成员国和地区借此成功实现了创新性的对地观测航天器编队。这 7 个成员分别是巴伐利亚州、佐治亚州、魁北克省、圣保罗州、山东省、上奥地利州和西开普省[②]。TIM 任务使用皮卫星编队,利用从不同观测方向进行摄影制图的优势,生成 3D 对地观测图像(图 3.6,参见彩图)。获得的数据将被融合处理,用于监测环境污染、农作物长势、关键基础设施和自然灾害(如森林火灾、火山活动、地震等)。

---

① 译者注:中国称 RLS 为"友好省州领导人峰会"。
② https://www.rls-sciences.org/small-satellites.html。

　　该任务目前处于实施阶段,计划发射时间是 2019 年①。相关科学挑战涉及航天器工程和科学数据处理:

　　(1) 与国际伙伴合作,开发模块化、高可靠的小卫星,形成网络化、相互合作的"智能"小卫星,自主运行,与地面站的交互最少。

　　(2) 摄影制图数据处理,用于生成 3D 图像,利用不同卫星上仪器之间的大基线距离,通过传感器数据融合方法获得更高的分辨率。

**图 3.6　3 颗"国际远程信息处理任务"(TIM)卫星通过摄影测量方法聚焦同一目标区域的 3D 图像**[图片来源:Zentrum fur Telematik (https://www.telematik-zentrum.de/)]

　　其中,编队所需的关键基本子系统有:姿态和轨道确定及控制系统、星间及卫星与地面站之间的通信系统,以及用于轨道控制和编队保持的电推进系统。编队

---

　　① 　译者注:TIM 的中国合作方是中国空间技术研究院山东航天电子技术研究所,截至 2020 年 3 月 22 日,尚未查到该计划发射信息。

的核心是 3 颗名为"远程地球观测任务"(Telematics earth Observation Mission, TOM)的巴伐利亚卫星。每颗卫星的加入都增加了编队的能力,仪器数量的增加也可以补充观测数据。在该国际团队中,合作伙伴之前的任务为相关领域的专业知识奠定了基础,使这一具有挑战性的皮卫星编队成为可能。一项具有挑战性和创新性的对地观测任务就是通过国际合作得以实现的。

### 3.5.1.2  QB50 任务案例研究

QB50 任务是来自世界 23 个不同国家的高校团队为实现科学目标而建立的立方星网络,这展示了由"新航天"产业(在学术界成长起来的初创公司)支持的国际高校间合作的潜力。该任务的主要目标包括让高校和研究中心更容易进入太空,进行热层测量、验证在轨新技术以及促进空间合作和科学教育。

QB50 任务对一个以前研究不多和无法进入的区域——低热层进行了综合测量。该任务监测热层不同气体分子和电学性质,以便更好地了解空间天气及其长期趋势,以及空间天气与气候变化的关系。QB50 任务提供的数据增强了大气模型,并提高了对空间天气如何干扰无线电通信和 GPS 信号的了解。这项研究有助于对可能损害地表电网和空间资产(即军用、商用和民用卫星)的强太阳事件进行风险评估。

该项目由 QB50 联盟协调管理,并得到了欧盟 FP7 计划的资助。各国空间机构目前不追求在低热层建立就位探测的航天器网络,因为按照工业标准建造的航天器网络成本极高,而且轨道寿命有限,这是不合算的。因此,中低热层的物理和化学现象只有通过使用成本非常低的卫星网络才能实现,而由高校建造的立方星是唯一现实的选择。

为了完成这项科学任务,由 3 所大学(伦敦大学、德累斯顿大学和奥斯陆大学)组成的联盟开发了 44 台高度小型化的仪器(图 3.7,参见彩图)。QB50 也让他们加深了对如何制造、部署和使用小型分布式传感器技术的理解,而这种技术的空间应用将越来越多。

QB50 星座的大部分卫星(36 颗立方星中的 28 颗)于 2017 年 4 月 18 日从卡纳维拉尔角(Cape Canaveral)的发射中心发射到国际空间站,一个月后被释放到太空(图 3.8,参见彩图)。2017 年 6 月进行了第 2 次发射,剩余的 8 颗立方星沿极地轨道运行。在发射的 36 颗立方星中,9 颗在抵达太空时失效或在发射后失去联系。对于 27 颗"幸存卫星"中的大多数,进行在轨测试也十分具有挑战性,只有 16 颗卫星每天都在产生有价值的科学数据。其中 IOD 立方星在发射后的 2 个月内成功完成任务:InflateSail 成功释放了牵引帆并重返大气层。这些立方星绕地球轨

图 3.7 欧盟 QB50 项目的三种类型科学传感器(从左到右):离子与中性质谱仪(INMS)、通量-Φ-探针实验(FIPEX)和多针朗缪尔探头(m-NLP)[图片来源:欧盟 QB50 项目科学小组(https://www.qb50.eu/)]

道运行期间轨道高度逐渐下降,最后在大气层中完全烧毁,寿命在 1~2 年。截至 2018 年 5 月(释放 1 年后),只有 6 颗立方星仍在完好运行。在漫长的下降过程中,这些卫星利用广泛分布的传感器网络进行了大量测量。最后一颗 QB50 立方星在释放 19 个月后于 2018 年 12 月重返大气层。

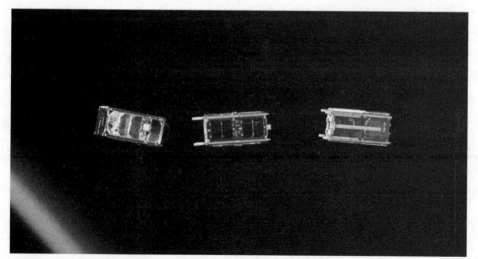

图 3.8 部署后的欧盟 QB50 卫星(图片来源:欧盟 QB50 卫星科学小组,冯·卡门研究所)

QB50 任务在实现其教育目的方面极为成功。QB50 立方星由大量年轻工程师设计、建造和操作,由他们所在大学的经验丰富的教职人员监督,并由 QB50 任务管理方通过评审和反馈意见进行指导,这些年轻的工程师将带着宝贵的实践经验走出校园。尽管科学目标的实现成败参半,但这种国际合作伙伴关系模式是未来大型星座任务的重要趋势。如果得到国家空间机构充足的资源支持,这种模型可以为科学研究提供巨大的潜力。

## 3.5.2 高等教育和共享经验教训

立方星科学任务能为高校学生和职业早期的航天专业人员提供实践培训机会,锻炼首席科学家的领导能力及科学、工程和项目管理技能。由于跨越多个科学和工程领域的开发过程十分复杂,所以高校学生和研究人员必须积极参与所有的开发阶段(有可能持续数年),从而获得能对科学做出重大贡献的技能。

小卫星项目是"项目式学习"(Project Based Learning,PBL)教学过程中的一个很好的案例。在这个过程中,学生在解决现实问题的同时可以积极学习并获得经验。与传统教学方法相比,科学、技术、工程和数学领域的主动学习方法显示出更好的学习成果。

高校和其他组织可以在多个领域共享经验教训。

### 3.5.2.1 共同课程开发和培训方法

"立方星设计"的课程因情况而异,许多高校没有正式的立方星课程。相反,立方星项目却往往被作为学生项目,并被纳入系统工程课程或航天器设计课程(NASEM,2016)。因此有必要制定一个共同的课程,让学生在不同高校学习类似的内容,以促进项目活动和国际层面的合作工作。

大学空间工程联盟[①](University Spacecraft Engineering Consortium,UNISEC)在学术界建立了一个全球教育网络,该网络整合了全球大学开展的活动与航天器设计活动,分享了与罐头卫星(CanSat)和立方星有关的教育材料,组织了联合工作组和会议,举办了航天器设计竞赛,并促进了支持子系统/部件交换的标准化工作。[②]

### 3.5.2.2 SpaceMaster 案例研究

欧盟"伊拉斯莫斯世界计划"[③]的"空间科学与工程专家"(SpaceMaster)是2005年发起的一个国际联合硕士项目,得到了 6 所欧洲大学的支持。在这个综合项目中,学生可以在 3 个国家进行学习,开设非常广泛的空间科学和工程

---

① http://www.unisec-global.org。

② http://unisec-europe.eu/standards/bus/。

③ 译者注:这一计划的英文名称为"Erasmus Mundus",其中"Erasmus"来自 15 世纪在欧洲各地修道院游学的一名荷兰籍学者的名字;"Mundus"是拉丁语中"世界"的意思。

专业①。

各大学为这些课程提供了各自的专业师资以涵盖大的航天器设计和空间环境这一跨学科领域。尤其是系统设计技术,这种技术对航空航天领域之外的工业应用也很有价值。学生可以选择学习空间物理方面的科学课程(重点是仪器和天文学,或者是大气和行星物理学),也可以学习聚焦航天器设计和任务实现的工程课程。

这种空间科技教育的国际性体现在不同的学习地点上:学生第一学期在德国乌兹堡,第二学期在瑞典基律纳;第二学年则根据所需专业,在 6 所合作的欧洲高校中选取一所进行学习。成功毕业的学生可以获得两所欧洲高校的双文凭(获得主要学分的两所学校)。学生群体也非常国际化:通常情况下从大约 600 份申请中选出 50 名学生,其中一半来自欧洲,另一半来自欧洲之外的国家。

另外一个特别的亮点是,学生有机会为撰写硕士论文而参加小卫星设计活动。在"伍兹堡大学实验卫星"(Universitat Wurzburg Experimental satellit,UWE)中,学生建造的"皮卫星"(pico-satellite)都已成功在轨运行,包括用于研究太空互联网的 UWE-1(2005 年发射)、研究卫星姿态的 UWE-2(2009 年发射)和解决卫星姿态控制的 UWE-3(2013 年发射)。UWE 计划的目标是逐步发展"皮卫星"编队飞行的全部技术。

### 3.5.2.3　分享从小卫星任务中获得的经验教训

美国国家科学院的报告(2016 年)证实,高校发射第 1 颗立方星的故障率通常高于第 3 颗或第 4 颗,这表明相关经验教训能够帮助他们建造更好的航天器。国际空间会议(如 COSPAR 和 IAC)为各团体提供了一个学习他人经验和分享经验教训的机会,因为高校学生和研究人员可以在会上与其他高校的同行以及空间机构和航天工业的高级代表会面。在讲座或研讨会上的非正式交流可以激发创新性对话,并带来新的倡议和合作。

### 3.5.2.4　合作研究需要复杂的架构或星座

高校合作型星座是一种低成本的选择,可以替代由工业界按照工业标准构建的星座。高校星座还适用于轨道寿命有限的科学任务,或在工业界中没有商业利益可图的科学任务。

---

① 　http://www.spacemaster.uni-wuerzburg.de,http://spacemaster.eu/。

### 3.5.2.5 共享资源和标准

在小卫星领域,高校间的国际合作使得他们在分担费用方面具有显著优势,因此有可能吸引传统空间机构以外的新赞助者进入空间科学领域。为了进一步促进国际合作,需要制定和扩展广泛采用的电气接口标准,如大学空间工程联盟-欧洲(University Space Engineering Consortium-Europe)标准,从而支持科学机构之间的部件交换。根据美国国家科学院的报告(2016年),一些关键的子系统已经可以购买商业货架产品,包括被标准化为立方星尺寸的电源板和通信系统等。可以购买的子系统和通用软件方面的进步使得科学立方星任务能够将主要资源和精力投入科学仪器的开发和科学研究任务本身。小卫星项目提供了在以下领域共享资源和定义标准的机会:

(1) 业余无线电爱好者分配频率使用手册(甚高频、超高频和S频段)。高校地面站能够在同一业余无线电频段与多个航天器通信。

(2) 共享频谱和通用接口,用于申请频率分配和协调。

充足的频谱分配不仅是一个技术问题,也是一个监管问题。通用接口将会简化高校申请频率分配的过程,并确保更有效地使用频谱。事实上,单个地面站可以重复使用同一频率在不同时段与多颗卫星通信。

(3) 开源的地面系统软件和星载软件的全球弹性标准。

(4) 高校的地面系统处于通信的任务时间通常不到3%,97%的时间地面站是空闲的。例如,在"相对论电子爆发强度、范围和动力学特性聚焦探测(Focused Investigations of Relativistic Electron Burst Intensity, Range, and Dynamics, FIREBIRD)任务"中,由于遥测系统的限制,地面只接收到0.5%的高速率数据。一个标准的地面站和航天器软件能让世界各地高校现有的地面站链接在一起,与其他高校的卫星通信,并通过互联网将任务数据传输到研究中心。该模型还可以进行扩展,成为"全球高校地面站网络",为所有参与的高校提供全球覆盖,并低成本从空间获取大量数据。许多关键的操作和科学任务都将受益于该网络不间断的覆盖,从而大幅度地增加了任务回报。

(5) 低功率标准发射机。这些发射器可以在小卫星功率有限情况下运行,同时有利于减少对其他卫星的干扰,从而增加空间总数据吞吐量。

(6) 标准释放器和共享发射。大多数立方星是作为大型火箭的搭载有效载荷而被释放到太空中的,并使用了运载适配器以便能够安装在运载火箭上。适配器既可以在太空中释放专门尺寸的立方星(通常长度可达3U),也可以用于释放不同尺寸立方星的组合(如 1U 和 2U 组合)。目前立方星的释放器仍然没有统一标准,这就意味着很多情况下卫星系统的设计取决于发射商的选择。例如,在 ISIS 公司的 QuadPack 立方星释放器中,卫星通道位于顶部面板;而 NanoRacks 立方星释放器的卫星通道则在侧面面板上。相应地,为电池充电或连接到星载计算机的卫星接口位置也会改变。如果能够确定一个可以容纳任何高达 12U 立方星并集成到任何发射装置上的标准释放器,则不仅具有成本效益,还能增加科学立方星可用轨道的多样性。

(7) 集中式系统工程。航空航天工程中最具挑战性的是如何设计空间任务中相互依赖的子系统(NASEM,2016 年),此系统能够确保空间任务的成功。适当的系统工程可以确保一个项目的所有可能方面都被考虑到,并集成一个整体。但不幸的是,与航空航天工业不同,几乎整个学术界都没有成熟的系统工程学科,也没有这个领域的知识和经验。因此,系统工程往往是高校项目中最薄弱之处,并可能导致空间科学任务的失败。因此,一个涉及多所高校的项目需要由专家管理的集中式系统工程,这些专家将为所有合作伙伴制定统一的实施标准。

(8) 国际项目的集中管理。在国际项目中,合作的高校应建立一个联合指导小组,以减少科学和技术领域的重复投资,将资源分配给各个领域中最合适的合作伙伴,为研究人员创建更简单的接口,使集成、测试和发射工作更具有一致性,并为供应商和发射商提供一个统一接口。

### 3.5.2.6　全球卫星运行教育网络(GENSO)案例研究

GENSO 是大学之间共享地面控制站资源的早期尝试,2007 年由 ESA 牵头的几个空间机构的教育计划提供支持。目前最活跃的全球卫星运行教育网络是前文提到的 UNISEC[①],约有 100 所开设航天器工程课程的国际高校参与其中。UNISEC 不仅支持全球地面站网络,还制定了电气接口标准[②](图 3.9,参见彩图),这些标准已经在欧洲和日本的几个立方星任务中成功实施,并为在国际合作任务中交换子系统和元件奠定了良好的基础。

近期一个类似的项目于 2014 年启动:"卫星网络开放地面站项目"(Satellite Networked Open Ground Station,SatNOGS)[③]是一个资源全球开放卫星地面站网络,主要观测和接收卫星信号,特别是近地轨道的立方星信号。

---

① http://www.unisec-global.org/。

② http://unisec-europe.eu/standards/bus/。

③ https://satnogs.org/。

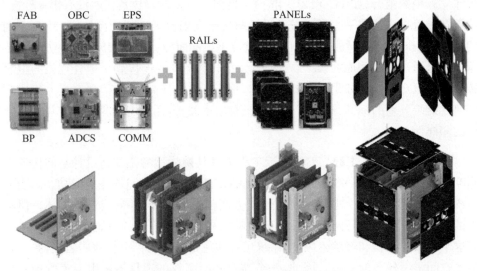

图 3.9 根据"大学空间工程联盟-欧洲"电气接口标准,通过子系统级别的模块化组件,可灵活组装的完整卫星,其中线束由背板代替(图片来源:Zentrum fur Telematik)

### 3.5.3 中学教育

"空间民主化"的发展越来越需要对能够完成简单复杂的工业任务的技术人员。虽然高校和研究生层次的高等科学教育得到了认可和加强,但也应该特别努力地促进中学阶段的科学和技术教育。小卫星的制造和组装将提供各种各样的技术工作,完成这些工作不一定需要硕士或博士学位。不仅仅是火箭科学家才能建造火箭!

中学教育水平非常适合引导年轻人进入职业学校或技术大学学习。这意味着科学家和工业界必须确定他们的需求,并尽可能花时间向老师和学生解释这个迷人的领域,使这些学校能够设置相应的课程,并为工业界提供所需的人员队伍。一个明显的社会效益是,这种方法可以更早地创造出提供工人工资、税收和相关经验的生产工作岗位。

瑞士国家机器人研究能力中心(National Centre for Competence in Research in Robotics)就采用了这种方法①。该中心提供了转移基金,使科学家和工程师能够在学术环境中从事他们所做的工作,并利用这些工作创建一个技术转移公司。

---

① https://nccr-robotics.ch/kids-teachers-parents/how-to-be-a-roboticist/。

科学家和工程师从实际项目中得到了支持,并使这些项目对公众或公司具有实际的应用价值,以此获得将产品推向市场所需的视野和广泛的技能。除了学术研究之外,许多机构都需要机器人专家,如医院、制造厂、环境服务机构及空间机构和工业界等。

如今教育工作者面临的问题之一是如何创造一个教学环境,在科学技术领域为学生提供有意义和有效的学习体验。"项目式学习"也被成功地应用于中学教育,这是一种动态的课堂教学方法,利用这种方法,高中生可以通过对现实问题的积极研究,来获得更深层次的知识。这种方法已被证明能够有效地提高学生的学习能力和兴趣。对"项目式学习"而言立方星项目非常有吸引力。中学教育工作者正在利用太空吸引年轻的一代,引导他们学习科学、技术、工程和数学,并培训他们成为未来的科学家或工程师。

### 3.5.3.1　鼓励年轻女性从事科学技术工作

美国人口普查局的报告显示,女性占美国劳动力的近一半,但仅占科学、技术、工程和数学工作者的24%。小卫星计划可作为平台,鼓励年轻女性从事航空航天和科学、技术、工程及数学工作。事实上,以色列有一项与公立宗教学校合作的非常成功的立方星计划,在公立宗教学校里,男生和女生被分在单性别教室中,研发团队可以完全由女生组成。事实证明,女生不仅拥有出色的研发能力,还具有领导和创业能力,而且她们的科技项目质量普遍优于同龄男生。

### 3.5.3.2　教师和学生的机构支持

近年来,ESA 一直在支持"欧洲空间教育资源办公室(European Space Education Resource Office,ESERO)项目",该项目计划建立联络/资源中心,配备教育专家,并纳入国家教育系统和网络。这些中心提供富有创意的材料来助力教师教学和学生学习,并支持教育推广活动。在一些计划中,高中生可以参加科学实验的研发,这些实验可以在各种微重力平台上飞行,如在探空气球上或被送到国际空间站(图 3.10,参见彩图)。

### 3.5.3.3　Duchifat 案例研究

Duchifat 是以色列中等教育系统的一个立方星计划,由 12～18 岁的学生参与。七年级时,他们开始学习基础科学课程;九年级时,学习成绩优异并表现出更强动力的学生会继续学习以立方星设计为重点的高级课程,并成为学校"卫星和空间实验室"的成员。每个学生团队由一名经验丰富的航空航天工程师带领,并负责卫星的一个子系统。团队的任务通常相当具体且明确,即使学生缺乏正规的工程

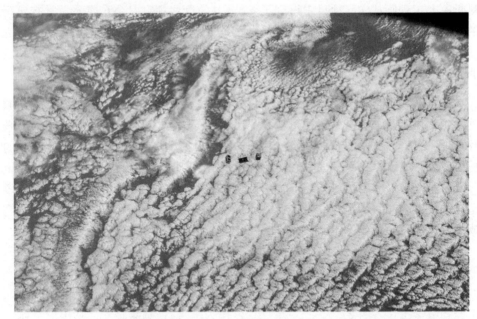

**图 3.10　从国际空间站上观看 3 颗分别来自西班牙、希腊和以色列的立方星**
**(图片来源:美国国家航空航天局)**

教育也能成功地完成任务。系统工程、集成和测试问题也由学生负责,但这些问题出现在后期阶段(通常是 12 年级,相当于我国的高三),那时学生更有经验,也成为了更年轻学生的导师和领导者。该计划的 2 颗立方星已成功进入太空:"Duchifat 1"卫星于 2014 年 6 月发射;"Duchifat 2"卫星又名 Hoopoe,作为欧盟 QB50 项目的一部分于 2017 年 5 月发射。截至 2018 年 5 月 2 颗卫星仍在完好运行。

此系列的另外 10 颗立方星,即 Duchifat 3~12,被用于生态应用和空间气象监测,目前正在以色列各地处于不同的开发阶段,计划于 2020 年发射。

"印度/以色列 75 颗星(INDIA/ISRAEL@75)计划"在创建学术界、工业界、政府和教育系统参与的新生态系统方面又向前迈出了一步。"INDIA/ISRAEL@75计划"是印度和以色列的一个合资计划,旨在 2022 年开发、建造并向太空发射 75 颗卫星,以庆祝两国独立 75 周年。这些卫星将由 75 所以色列和印度的高中及大学制造,并形成一个覆盖地球表面的星座。此外,这些相对简单的立方星(大小在 1U 和 3U 之间)将能够从地面上传送算法,作为科学实验和测试未来技术的平台。该星座将由两国设立的地面站控制和指挥。在这个新颖的生态系统(学术界—工业界—教育界—政府)中,教学人员将主要以数学、物理和计算机科学的教师及研究人员为基础,但也包括来自以色列和印度航空航天相关学科的经验丰富的工程师和专家。教学人员将指导各年龄段和水平的学生(从高中生到博士生)。该计划以两国在制造立方星方面的传统和经验为基础,并得到了两国政府的支持。

## 3.5.4　促进国际合作

不同国家之间还有其他几个现有的合作框架。例如,BIRDS 卫星项目①是一个由日本支持的跨界跨学科项目,参与国包括加纳、蒙古、尼日利亚和孟加拉国。在这个为期两年的项目中,各国学生可以设计、开发和运行 5 套相同的 1U 立方星。英国于 2015 年启动了"国际伙伴关系项目"(International Partnership Program,IPP)②,旨在为不发达国家和发展中经济体带来可持续的经济或社会效益。

目前涉及国际合作的大型航天器的遴选模式不太适合小卫星,这是因为所涉合作伙伴的计划框架不同,诸如 Solar C 和"国际太阳极轨道任务"(International Solar Polar Mission,ISPM)等历史案例都说明了这一困难。启动欧盟 QB50 项目则是朝着正确方向迈出的一步。

COSPAR 与各合作伙伴举办能力建设培训班③的传统由来已久,目的是传播 COSPAR 感兴趣领域的实用知识,并在科学家之间建立长久联系。这可进一步发展为一个与美国《十年调查》类似的活动,在国际或全球层面进行。COSPAR 在这一进程中发挥主导作用。

> **研究发现 3.11**　COSPAR 作为空间领域第一个也是最具权威的国际组织,能够很好地支持国际社会建立和协调全球(乃至深空)小卫星网络的基础设施或工具,任何人都可以通过明确定义的格式和接口,为该网络做出贡献,从而创建一个由所有贡献者组成的虚拟星座,这将远远超过单个个体的自身能力之和。

---

① http://www.birds-project.com。

② https://www.gov.uk/government/collections/international-partnership-programme。

③ https://cosparhq.cnes.frlevents/cb-workshops/。

# 第4章

# 空间科学小卫星发展建议

根据本书前三章的研究结果,本章给出如下5条建议作为结论,这5条建议的对象分别是科学界、航天工业界、航天机构、政策制定者以及COSPAR。

### 1. 对科学界

整个科学界应该认同小卫星的价值,并寻找机会利用小卫星工业界的新发展形势。空间科学各个领域都可能从更小尺寸、更低成本和更快迭代周期的小卫星发展中获益,尤其是小国家的空间科学界会受益于对小卫星的投资。

### 2. 对航天工业界

卫星开发企业应该主动寻求与科学家、高校和大型政府机构的合作机会。具体而言,合作机会可能涵盖数据共享协议、出售商业航天器上可以搭载有效载荷的空间等。目前,开放数据对实现科学目标而言有非常高的价值,因此商业公司应该对科学使用的免费、完整和开放的数据协议持开放态度,这种合作关系也有助于相关人力资源的发展。

### 3. 对航天机构

大型航天机构应该根据小卫星项目的体量制定合适的步骤和程序。各国航天机构应该探索新方式为科学、应用和技术验证类小卫星提供发展机遇和充足的发射窗口。此外,各国航天机构应该充分利用商业数据和商业航天设施来进行科学研究,并遵循数据开放政策。最后,各国航天机构应该携手制定小卫星发展长期路线图,以明确未来优先发展的国际合作小卫星任务。

### 4．对政策制定者

为了确保空间科学小卫星任务的成功,科学界需要政策制定者的如下支持：① 确保频谱的充分可用,采取轨道碎片减缓和整治方案,以及提供经济上可负担的发射成本和其他基础设施服务；② 确保制定的出口管制指南易于理解和解释,并做好国家安全和科学利益之间的平衡；③ 针对小卫星频谱获取、轨道机动性、可跟踪性和寿终报废处置等方面的国家和国际法规开展教育和指导。

### 5．对国际空间研究委员会(COSPAR)

国际团队可以共同为类似于欧盟 QB50 这样的模块化国际小卫星星座任务确立科学目标和规则,而 COSPAR 在这个过程中应起到促进作用。通过 1957～1958 年"国际地球物理年"(IGY)这样的活动,参与者可以就小卫星发展的基本规则达成一致,而各国航天机构或相关机构代表应该从一开始就参与进来。资金来源于各成员国,甚至可以来自私人实体或基金会。COSPAR 并不提供资助,而是扮演忠实的协调角色,并规定这些国际团队在提出任务建议时必须遵守的准则。

通过这种国际合作所产生的成果对所有任务参与方来说都是十分有价值的,并且总体价值高于各国"单打独斗"所产生的价值之和。COSPAR 会以开放的姿态为搭建"社区公民科学"创造典范。能够参与全球项目并获得全部项目数据将是对参与者的激励。最后提出的这条建议是促进真正宏大想法付诸实践的一种途径,它们包括本书所谈到的 4 个未来愿景以及其他目前还未想象到的宏大想法。

# 第 5 章

## 昨天和今天

在太空时代的早期，做出决策的速度快，空间项目的发起和完成周期也快。苏联于 1957 年 10 月 4 日发射了人类第一颗人造卫星；美国的 NASA 于次年 7 月成立，距离苏联卫星发射不足一年；"阿波罗（Apollo）计划"始于 1961 年。1962 年，NASA 咨询委员会（Advisory Council）向美国国家科学院空间研究委员会（Space Studies Board，NAS）提出需求：为空间科学制定一系列高优先级的任务目标。第一个"轨道太阳天文台"（Orbiting Solar Observatory，OSO 1）于 1962 年 3 月发射。

1967 年 11 月 9 日，巨大的"土星五号"（Saturn V）无人测试火箭发射；第一次土星五号载人发射发生于 1968 年 10 月 11 日；人类首次载人探月任务始于 1968 年 12 月 21 日，1969 年 7 月 20 日实现了人类登陆月球。在"阿波罗计划"期间，轨道太阳天文台 3 号、4 号、5 号、6 号和"天空实验室"（Skylab）相继发射。

当时没有电子邮件、Excel、PowerPoint，甚至没有具备相当于现在低端智能手机性能的电脑。"阿波罗"计划就是在这种条件下完成的。那时的通信手段是信件、电话或传真。公平地说，当时的财政环境与现在相比也非常不同，在冷战时期，任务预算成本基本不是问题。另外，现在的责任制度也比过去严格得多。

现在，NASA 的"帕克太阳探测器"（Parker Solar Probe）于 2018 年 8 月发射，ESA 和 NASA 的"太阳轨道器"（Solar Orbiter）正处于最后的测试阶段，将于 2020

年发射。① 太阳探测器(Solar Probe)是 2002 年《十年调查》推荐的太阳和日球层任务。而且在太阳探测器进入推荐名单之前,相关项目计划和论证已经进行了若干年。ESA 的太阳轨道器任务始于 1994 年。这两个项目都经历了至少 20 年的积极开发和研制,这个周期是从决定登月到实现载人登月任务周期的两倍。

因为都要接近太阳,所以这两项任务在技术上都很复杂,但很难说这些技术挑战比"阿波罗计划"中的技术挑战更大。此外,20 世纪 70 年代发射的"太阳神号任务"(Helios Mission)的两颗探测器和如今的太阳轨道器距离太阳的位置是相当的。作为机器,太阳轨道器和太阳探测器都不需要载人航天任务所需要的严格监管。

太阳轨道器和太阳探测器两项任务并不是孤例。JWST 进入推荐名单的日期比它们还要早几年,但目前最早的计划发射日期是 2021 年 3 月,而前提是 JWST 最终测试能按计划进行。②

NASA 和 ESA 都已认识到空间科学任务周期太长这一问题,并已制定了旨在缩短卫星开发研制周期的计划。NASA 设置了"小型探索者(SMEX)计划"和"地球探险(Earth Ventures)计划",ESA 也创建了 S 级任务(小卫星)。这些项目的执行周期约为 7 年,在计算该项目的周期时,结束时间是收到第一批科学数据的时间。这些任务的计划节点也需要保持研发资金的到位率。

一项空间任务的能力取决于计算机集成电路和存储器的性能。几十年来,集成电路的集成度和性能每 18 个月翻一番。这导致目前在轨任务受到计算机性能的限制,这些计算机要执行现代软件系统的指令,而软件的运行又需要快速处理器和大内存。现在,要获得 2TB 的固态内存只需要在网络上点击几下,花费几百美元即可,这比目前在轨的 SMEX 卫星上的内存还要大很多。

目前,这一代空间任务在发射时技术上已经落后,这一事实凸显了在资金、人力技术资源和科学资源使用方面的效率较低。在卫星研制阶段产生的技术进步甚至会让一些卫星任务的科学目标过时。这是可以理解的,因为任务的发射频率很低,机会很少,只有在技术和科学上被认为风险最低的任务才能经过漫长的申请过程,通过一扇扇评估和遴选的大门。

---

① 译者注:太阳轨道器是 ESA 和 NASA 的联合任务,已于 2020 年 2 月 10 日成功发射,它将首次给太阳南北两极拍"正面照",揭示太阳磁场的奥秘。

② 译者注:NASA 于 2020 年 7 月宣布,"詹姆斯·韦伯空间望远镜"发射日期推迟至 2021 年 10 月 31 日。在过去十年中,备受期待的韦伯望远镜遭受了一系列成本超支和计划延误。自 2009 年以来,预算几乎增加了一倍,发射日期总共被推迟了近 7 年。

但是,现在这种情况正在改观。新一代商用火箭的发射成本更低,这样就降低了占据任务成本比例最大的发射成本。因此,我们可以期待不久就会有更多国家在空间科学领域扮演更重要的角色。目前,中国、印度、澳大利亚、韩国和阿拉伯联合酋长国等国家政府正在大力发展空间科学项目,并和 NASA、ESA、俄罗斯国家航天集团公司(ROSCOSMOS)和日本 JAXA/宇宙科学研究所(ISAS)一起促进空间科学发展。

1999 年,来自加州理工大学和斯坦福大学的两位教授为学生使用的空间实验的小型卫星写了一份指导手册。这颗小卫星是 10 cm 立方体模块化设计的产物:立方星。当时,两位教授策划了鼓励学生参加空间实验的项目。立方星在过去十几年中不断得到发展,现在已经开始携带重要的科学载荷。三个因素的结合才产生了这一变化:新的低成本发射机会;重量轻且低功耗的 CPU;经过验证能够耐受近地轨道空间环境的大容量存储芯片。立方星及更大的纳卫星和小卫星正在提供新的机遇,可以使空间科学研究速度更快、成本更低。此外,它们还可以在星座任务中互动,这很有可能会开创空间科学发展的新时代。

空间活动不再是大型航天局的独角戏。在过去的几年里,风险投资人越来越认识到,从太空对地球进行监测所获得的数据很有市场价值。商业项目已经发射了数百颗立方星和小卫星。小卫星项目的成本虽然很高,但对高校来说,拥有自己卫星任务的成本还能承受。在日本、德国、以色列、意大利、法国、英国、瑞士、韩国等国家,高校已经在开展自己的卫星项目。2016 年,美国国家科学院发布了一份报告《用立方星实现空间科学》,而本书报告了小卫星用于空间科学的现状和未来前景。

正如本书开篇所说,我们最终的追求是:由科学家组成的不同国际科学团队努力实现具有创新意义和具有深远影响的科学目标。本书提供了一些通过小卫星实现这一追求的可能途径。对于空间科学卫星任务,如下变化可能成为常态:质量缩减为几十或者几百千克,而不是几吨;开发研制周期缩减为几年,而不是几十年;总成本下降到几千万美元而不是几十亿美元。通过小卫星星座任务,小卫星的潜力会得到进一步扩大,它不仅能够提供多点观测,而且能够增加容错能力,因为单个网络节点的故障对整个网络的影响很小。一项由数千颗地球观测卫星组网而成的任务可能会带来具有巨大科学和社会影响的应用。通过小卫星集群探测太阳系中的一个独特目标天体(如 1P/哈雷彗星),集群中的每颗小卫星可以执行不同的观测任务,由不同的航天局研制,这样获得的回报可能超过任何一个单独实施的任务。对空间天文领域来说,小卫星集群优势更加明显,因为很显然几乎不可能将比 JWST 更大的航天器送入太空。要想在一个人的生命周期内实现星际探索任务、到达太阳系外的某颗恒星,小卫星必须在星际任务似乎还遥不可及时就变得足够小。上述未来任务以及类似愿景目标所面临的技术挑战是巨大的,但与掌握技术同样重要的是建立国际合作的新方式和新方法。涉及国际合作的参与者包括高校和科研机构所代表的科学界、空间机构/航天局、航天工业界,以及政府和国际组织所代表的政策制定者。希望 COSPAR 能在国际合作过程中起到积极和重要的作用。

# 致　　谢

本项研究获得了国际空间科学研究所(ISSI)和 COSPAR 的支持和资助。其中 ISSI 主办和资助了两次论坛,COSPAR 为部分与会者提供了旅费支持。

# Small Satellites for Space Science
## —A COSPAR Scientific Roadmap

## Abstract

This is a COSPAR roadmap to advance the frontiers of science through innovation and international collaboration using small satellites. The world of small satellites is evolving quickly and an opportunity exists to leverage these developments to make scientific progress. In particular, the increasing availability of low-cost launch and commercially available hardware provides an opportunity to reduce the overall cost of science missions. This in turn should increase flight rates and encourage scientists to propose more innovative concepts, leading to scientific breakthroughs. Moreover, new computer technologies and methods are changing the way data are acquired, managed, and processed. The large data sets enabled by small satellites will require a new paradigm for scientific data analysis. In this roadmap we provide several examples of long-term scientific visions that could be enabled by the small satellite revolution. For the purpose of this report, the term "small satellite" is somewhat arbitrarily defined as a spacecraft with an upper mass limit in the range of a few hundred kilograms. The mass limit is less important than the processes used to build and launch these satellites. The goal of this roadmap is to encourage the space science community to leverage developments in the small satellite industry in order to increase flight rates, and change the way small science satellites are built and managed. Five recommendations are made; one each to the science community, to space industry, to space agencies, to policy makers, and finally, to COSPAR.

Keywords: Small satellites; Space science

## 0　Executive summary

In early 2017, an international study team of science and engineering leaders under the auspices of COSPAR embarked on a 2-year activity to develop an international scientific roadmap on Small Satellites for Space Science (4S)①. For the purposes of this study, the committee defined "small satellites" to have an upper mass limit in the range of a few hundred kilograms. The mass limit is less important than the processes used to build and launch these satellites. Because CubeSats have played a critical role in the small satellite revolution, significant discussion on CubeSats and CubeSat technology-enabled small satellites is included. CubeSats are small satellites built in increments of 10 cm cubes (1 cube is called 1U or "unit", two 10 cm cubes together are known as 2U, and so on). This report is motivated by recent progress and results summarized in an article in Space Research Today (Zurbuchen et al. , 2016) and a study by the US National Academies of Sciences, Engineering, and Medicine (NASEM, 2016). ② The committee on the roadmap for small satellites for space science was tasked with addressing six specific questions:

(1) What are the status and use of small satellites, in particular CubeSats, for science, their technological capabilities, and their key successes to date?

(2) What is the scientific potential of small satellites both as stand-alone targeted missions, but also as secondary payloads, and as constellations and swarms?

(3) What is the role of participating agencies and industry in developing standardized approaches to the development of spacecraft (hardware and software), and also ground-systems, etc. that enables this science?

(4) What are the policies that support the growth of the number and types of CubeSats and CubeSat technology enabled small satellites, related to communications and frequency allocation, orbital debris, and launch vehicles?

(5) What are successful models for international collaboration between teams developing and operating small missions, and how are data being shared and preserved for the future?

(6) How can participating universities and international organizations learn from each other to share lessons learned and drive international collaborations in this rapidly moving field?

The COSPAR roadmap was developed by a study team that covers the broad range of scientific disciplines that use space-based observations, including Earth Science, solar and space physics, planetary science, and astronomy. The study team includes scientists, engineers, and policy experts working in universities, public research

---

① A list of acronyms is provided in the Appendix A.

② Available at https://www. nap. edu/catalog/23503/achieving-science-with-cubesats-thinking-inside-the-box.

institutions, and industry. The report aims to address the above questions in a way that is of value to space agencies internationally and their supporting governments, as well as non-profits and other private sector organizations that would be interested in promoting global SmallSat-based missions. Moreover, we hope to encourage the science community to leverage developments in the small satellite industry, and to pursue partnerships with industry and each other internationally, in order to increase flight rates and change the way small science satellites are built and managed, ultimately leading to significant advances in space science.

This roadmap document is in some ways similar to a trail map billboard posted in many mountain resorts for hiking (or skiing), depicting the village in the foreground as our neighborhood, some high and forbidding peaks in the distance representing the visionary goals we long to reach, and a system of trails leading us there across the mountainous hazards. Thus our roadmap is structured into three main parts: Section 1, "Our Neighborhood", provides an overview of the history and current landscape, including current status of technology and near-term scientific potential of small satellites and CubeSats. Section 2, "Visions for the Future", describes examples of the type of science missions that could be achieved in the more distant future—a decade and beyond. The science concepts and priorities of the future should ultimately come from the science community, however, the visions outlined in Section 2 serve to focus the discussion. In Section 3, "Obstacles and Ways to Overcome Them", we describe some institutional roadblocks and means to overcome them. In particular, the roles of agencies, industry, policies, and models of international collaboration and exchange are discussed.

The ultimate destination is a world in which international teams of scientists pursue novel and far-reaching goals. This roadmap provides some possible paths to reach such goals. We articulate a number of findings which are distributed throughout the sections. These findings lead to five recommendations:

### Recommendation 1—To the science community:

The science community as a whole should acknowledge the usefulness of small satellites and look for opportunities to leverage developments in the small satellite industry. All branches of space science can potentially benefit from the smaller envelope, the associated lower cost, and higher repeat rate. Scientific communities from small countries in particular may benefit from investing their budgets in small satellites.

### Recommendation 2—To space industry:

Satellite developers should seek out opportunities to partner with individual scientists and universities as well as larger government agencies. This might include data sharing arrangements, selling space on commercial spacecraft for scientific instruments, etc. Currently, publicly available operational data is very valuable for achieving science objectives. Commercial entities should be open to agreements that would continue to make such data available under a free, full, and open data policy for scientific use. Such partnerships can also contribute to workforce development.

### Recommendation 3—To space agencies:

Large space agencies should adopt procedures and processes that are appropriate to the scale of the project. Agencies should find new ways to provide opportunities for science, applications, and technology demonstrations based on small satellites and with ambitious time to launch. Agencies should additionally take advantage of commercial data or commercial infrastructure for doing science in a manner that preserves open data policies. Finally, space agencies should work together to create long-term roadmaps that outline priorities for future international missions involving small satellites.

### Recommendation 4—To policy makers:

In order for scientific small satellites to succeed, the scientific community needs support from policy makers to: ① ensure adequate access to spectrum, orbital debris mitigation and remediation options, and affordable launch and other infrastructure services; ② ensure that export control guidelines are easier to understand and interpret, and establish a balance between national security and scientific interests; ③ provide education and guidance on national and international regulations related to access to spectrum, maneuverability, trackability, and end-of-life disposal of small satellites.

### Recommendation 5—To COSPAR:

COSPAR should facilitate a process whereby International Teams can come together to define science goals and rules for a QB50-like, modular, international small satellite constellation. Through an activity like the International Geophysical Year in 1957-1958 (IGY), participants would agree on the ground rules. Agency or national representatives should be involved from the beginning. The funding would come from the individual participating member states for their individual contributions, or even from private entities or foundations. The role of COSPAR is one of an honest broker, coordinating, not funding. COSPAR should define criteria that must be met by these international teams for proposing.

The results of an international effort to build small satellite constellations would be valuable for all of the participants, and would be more valuable than the individual parts. Such a large-scale effort would enable the pursuit of visionary goals, and ultimately lead to both technological and scientific breakthroughs. Small satellites enable new models of international collaboration, with involvement by many more nations, in worldwide, ambitious projects. COSPAR is in a position to help foster this international collaboration, creating a precedent for setting up community science in a very open way. Our final recommendation is a means to facilitate progress towards really big ideas such as our four visions for the future or other ideas that we haven't yet imagined.

# 1　Our neighborhood

This section provides an overview of the small satellite landscape (Section 1. 1) and near-term scientific potential (Section 1. 2) of small satellites.

## 1.1　History and current status of small satellites and CubeSats

The small satellite industry is changing very rapidly. Here, we present a brief history of small scientific satellites and a brief overview of the industry at a snapshot in time. A more general review of modern small satellites is provided in Sweeting (2018).

### 1. 1. 1　Traditional small satellites for science

Small satellites in the mass range above approximately 100 kg have aptly demonstrated their utility for scientific missions, and have been essential contributors to Space Science knowledge for decades, specifically in the subdisciplines of Heliophysics[1], Astrophysics, and Earth Sciences.

In the U. S. , most scientific small satellites are supported by the NASA Explorers Program which provides flight opportunities for scientific investigations in Astrophysics and Heliophysics. Since it's beginning, with Explorer-1 in 1958, the program has supported more than 70 U. S. and cooperative international scientific space missions (more than 90 individual satellites). "Explorer satellites have made impressive

---

① Heliophysics encompasses studies of the space environment in the interplanetary medium including study of the Sun, heliosphere, geospace, and the interaction between the solar system and interstellar space.

discoveries: Earth's magnetosphere and the shape of its gravity field; the solar wind; properties of micrometeoroids raining down on the earth; much about ultraviolet, cosmic, and X-rays from the solar system and the universe beyond; ionospheric physics; solar plasma; energetic particles; and atmospheric physics. These missions have also investigated air density, radio astronomy, geodesy, and gamma ray astronomy. Some Explorer spacecraft have even traveled to other planets, and some have monitored the Sun. "[1] The early Explorers, launched between 1958 and 1962, massed less than 50 kg. Capabilities increased rapidly, but so did the mass. By 1989, the 66th Explorer mission, Cosmic Background Explorer (COBE) had a dry mass of 1408 kg. To address the rapid increase in mass and the resulting increase in cost and decrease in launch cadence, in 1988, NASA started the modern Explorers Program which enabled the development of small sciencecraft with masses in the range of ~60 – 350 kg. Missions within this mass range encompass a total of about 17 satellites, including satellites from the University Explorer line (UNEX), the Small Explorer line (SMEX) and a single 5-satellite Medium Class Explorer (MIDEX). The launch mass of ten SMEX missions is shown in Table 1.

The Explorers program has been extremely successful in terms of scientific return. However, it falls short with respect to increasing the launch cadence. Between 1958 and 1980, 62 Explorers were launched (2. 82/year), while between 1980 (the start of the Shuttle era) and 2018, only 33 were launched (0. 87/year), more than a factor of 3 decrease. The time between the last two solar physics mission launches is 11. 5 years: RHESSI in February 2002 to IRIS in June 2013. IRIS was the last SMEX mission launched, now more than five years ago. Increases in management oversight have likely contributed to an increase in development time. The high cost of launch may also be a driving factor in the launch cadence decrease. For reference, NASA's most recent Explorers mission, TESS (A MIDEX with launch mass 362 kg) cost $200 million (Wall, 2018), not including launch cost which was an additional $87 million,[2] more than 30% of total mission cost.

NASA's Earth Ventures line of missions was recently established to provide opportunities for small satellite missions in Earth Sciences. The first Venture-class

---

[1]  History of the Explorers Program, retrieved on May 27, 2018, from https:// explorers. gsfc. nasa. gov/history. html.

[2]  CONTRACT RELEASE C14-049, retrieved from https://www. nasa. gov/press/2014/ december/nasa-awards-launch-services-contract-for-transiting-exoplanet-survey-satellite on 2018, June 26.

satellite mission, CYGNSS (Cyclone Global Navigation Satellite System), measures ocean winds, and is demonstrating the utility of satellite constellations for Earth Science. Each of the eight simultaneously-operating satellites has a mass of ～28 kg. CYGNSS was selected for development in 2012① and launched in December 2016. ②

Table 1 　Launch mass of selected small satellites from NASA's Modern Explorers Program (Data taken from https://nssdc. gsfc. nasa. gov/multi/explorer. html, with exception of AIM and NuSTAR which were taken from the mission websites. )

| MISSION | MASS (kg) |
|---------|-----------|
| SAMPEX  | 158 |
| FAST    | 187 |
| TRACE   | 250 |
| SWAS    | 288 |
| RHESSI  | 230 |
| GALEX   | 280 |
| AIM     | 197 |
| IBEX    | 80 |
| NuSTAR  | 350 |
| IRIS    | 200 |

European activities on small satellites have generally been supported at different levels by national programs, by the European Union FP7 and Horizon 2020 Space program, and by ESA, the European Space Agency. A short history of ESA small satellites is given in Dale and Whitcomb (1994), and briefly summarized here. Small missions in Europe were first considered in association with the Space Science: Horizon 2000 strategic plan in 1985. At that time, the Cluster mission was being developed and procurement rules similar to those used for the NASA's AMPTE mission were considered. However, it was concluded that, "the changes needed to apply a similar 'small satellite' approach to Cluster were too wide-ranging and the project proceeded along more classical lines". In 1990, ESA issued a "Call for Ideas" for small

---

① 　https://www. nasa. gov/home/hqnews/2012/jun/HQ_12-203_Earth_Venture_Space_System_CYGNSS. html.

② 　https://www. nasa. gov/image-feature/cygnss-satellites-launched-aboard-pegasus-xl-rocket.

missions; 52 proposals were received and evaluated. Two missions were selected for further study: SOLID, a mission to measure solar oblateness, irradiance periodicities and diameter variations, and CUBE, a mission to survey the cosmic ultraviolet background. In November 1992 a specific request for small-mission proposals was released. Although some 13 small-mission proposals were evaluated, none were recommended for further study. The report concluded that, "While the ESA Science Programme Directorate has, as yet, no fixed policy on the practicality and potential for the introduction of a small satellite programme, there is a recognised need to reduce the overall costs of missions, which would allow more flight opportunities and a small spacecraft programme." The report also pointed out that smaller nations might not have the infrastructure needed to design, build and launch a small mission, so, ESA could potentially provide flight opportunities or act as a "go-between" for national programmes. In fact, the Belgian PROBA series of microsatellites was funded through ESA's small satellite program. The 94 kg PROBA-1 spacecraft, launched in 2001, is the longest flying Earth observing mission. [1] PROBA-2 and PROBA-V have also been operating for a number of years, collecting data on solar activity and vegetation/land-use, respectively. ESA's first lunar mission, SMART-1, weighed in at just over 350 kg. The spacecraft was launched in September 2003 as a rideshare to GTO and also provided a test of solar electric propulsion. [2]

More recently, in 2012, ESA announced that it will fund a new regular class of small missions, "S-class", in part to provide smaller member states the opportunity to lead missions. [3] Approximately 70 letters of Intent were received in response to the first call for proposals, [4] demonstrating a significant interest in small satellites. One of these missions was selected in 2012, CHEOPS (CHaracterising ExOPlanets Satellite), and is scheduled to launch in 2019. The second S-class mission, SMILE, is being developed jointly with the Chinese Academy of Sciences (CAS) to study solar wind-magnetosphere-ionosphere interactions. Despite being an S-class mission, the SMILE spacecraft is not really small; carrying four major instruments, the spacecraft has a dry mass of 652 kg and 1960 kg with propellant included (Raab et al. , 2016). Thus,

---

[1]  https://www. esa. int/Our_Activities/Observing_the_Earth/Proba-1.

[2]  http://sci. esa. int/smart-1/38890-smart-1-mission-to-the-moon-status-first-results-and-goals/.

[3]  https://www. bbc. com/news/science-environment-17335339.

[4]  http://sci. esa. int/cosmic-vision/50265-received-letters-of-intent/, retrieved on July 8, 2018.

similar to the NASA Explorers program, ESA opportunities for satellites in the $\sim 100$ kg range are limited, and mission development times approach ten years.

ESA's involvement in scientific microsatellites has been sporadic, and it is the ESA member states, often on a national basis, that have provided a growth of launches in the small/micro satellite class, opening the space sector and making it affordable to new international players. From 1988 to 2016, ESA launched only 6 satellites with mass below 200 kg, while individual ESA member states launched a total of 131 small satellites (Fig. 1. 1). [1] For example, in the early 90s, a Swedish-German mission, Freja (214 kg), [2] was launched to study the aurora. Other early players include Denmark which successfully conceived, designed, built and operated the geomagnetic mapping mission "Ørsted" (61 kg), launched in 1999. Ørsted provided information about Earth's dynamo (Hulot et al. , 2002), improved our understanding of ionospheric and magnetospheric current systems (Papitashvili et al. , 2002), and provided data that has been used as the source of the IGRF (International Geomagnetic Reference Field) model for half a decade. The French space agency, CNES, has launched a number of small satellites and developed the $\sim 100$ kg myriade platform which has been used for both Earth Science and military missions, beginning with Demeter, launched in 2004, and most recently, Taranis, which is slated for launch in 2019. [3] More recently, the Italian Space Agency (ASI) started a small satellites initiative, PLATiNO, which aims to establish a national capability for scientific and other missions through development of a multi-purpose small satellite platform. [4]

Russia has a long history of launching small satellites beginning with Sputnik-1 in 1957. Universities in Russia have been particularly active in developing small satellites for science. For example, Tatyana-2 is an international microsatellite ($\sim 100$ kg) mission led by Moscow State University, launched as a secondary payload in 2009 to study transient luminous events in Earth's atmosphere.

In some countries, small satellites are being developed primarily for industrial or

---

[1] It should be noted that some of these may have received funding through ESA programs.

[2] https://nssdc. gsfc. nasa. gov/nmc/spacecraftDisplay. do? id=1992-064A.

[3] https://myriade. cnes. fr/en/home-49.

[4] A Multi-Purpose PLATiNO SmallSat is the Plan by SITAEL and the Italian Space Agency, SatNews, December 19, 2017, http://www. sat-news. com/story. php? number = 275392102.

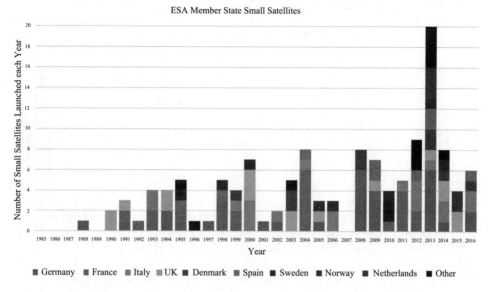

**Fig. 1. 1** **Small Satellites (<200 kg) launched by ESA Member States from 1985 to 2016.**
**For comparison there were 6 ESA small satellites launched during the same**
**timeframe. Data collected for Lal et al. (2017), provided courtesy of B. Lal**

operational use rather than science. For example, in Japan, the science agency JAXA has been pursuing tech demo microsatellites such as PROCYON, but otherwise mostly builds large satellite missions. Small satellites are primarily viewed as means for industrial development and a way to improve life (e. g. , using small satellites for Earth observation applications such as tsunami prediction). For example, Hokkaido and Tohoku Universities recently initiated a program to launch 50 microsatellites by 2020 for natural disaster monitoring. ① The program has participation from a number of countries in the region; 50 kg Diwata-1 was the first satellite built fully by the Philippines, and was deployed from the ISS in April 2016. ②In the last five years, the majority of small satellites were launched by the U.S. , but an increasing number of nations are developing small satellites (Fig. 1. 2). In particular, the number of small satellites launched by China is significant, though most of these are military or industrial use. For a comprehensive assessment of current international small satellite

---

① http://www. satnews. com/story. php? number=900912903. Retrieved on March 4, 2019.

② https://www. rappler. com/science-nature/earth-space/130956-diwata-microsatellite-deployment-space. Retrieved on May 27, 2019.

programs see Lal et al. (2017), Appendix E.

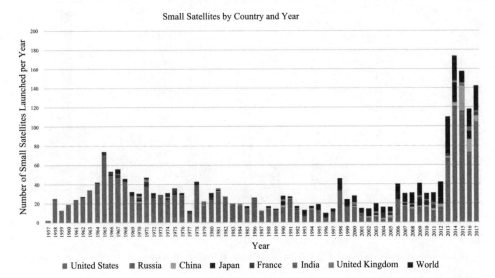

Fig. 1. 2  Small (<200 kg) satellite launches per year from 1957 to 2017 by country. Figure reproduced from Lal et al., (2017), "Global Trends in Small Satellites", figure E-2, with permission from the Institute for Defense Analyses

Worldwide, the number of satellites used for science is a tiny fraction of the total number of small satellites launched; the majority of small satellites are used for remote sensing or technology development (Fig. 1. 3). Small satellites in the mass range less than 200 kg offer an enormous potential for science, discussed further below.

## 1. 1. 2  CubeSats

Over the past decade and a half, a new class of satellites, called CubeSats, with masses between 1 and 12 kg has exploded upon the scene. Employed initially for hands-on technical training of college and university students (e. g. , SwissCube launched in 2009, see Noca et al. , 2009), approximately 1030 of these CubeSats have been launched through the end of 2018. [1]

CubeSats, so-called because the initial version of these satellites was in the shape of a cube measuring 10 cm × 10 cm × 10 cm (known as 1U), are a class of nanosatellites typically launched and deployed into space from a standardized container

---

[1]  Erik Kulu, Nanosatellite & CubeSat Database, https://www. nanosats. eu/.

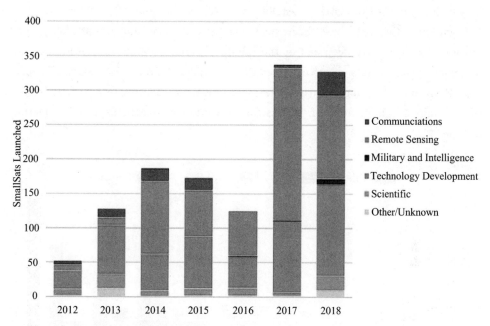

**Fig. 1. 3    Number of SmallSats by use. Image credit: Bryce Space and Technology (https://brycetech.com/reports.html)**

or canister, most frequently hitching a ride into space as a secondary or tertiary payload along with one or more larger spacecraft. The standardization of canisterized "CubeSats" allows for smaller and larger form factors consisting of fractional or multiple Us: 0.5U, 1U, 1.5U, 3U, and 6U CubeSats massing up to 12 kg have been launched.

For this roadmap, it is instructive to briefly review the history of the explosive growth of CubeSats, with focus on scientific applications. Fig. 1. 4 illustrates the annual launch rate of all CubeSats (and pre-CubeSat nanosatellites) launched worldwide between 2000 and 2018. The more than 450 CubeSats launched by Spire and Planet as elements of their respective constellations beginning in 2014 are not included in the chart since their numbers completely dwarf the others. The majority of CubeSats launched were targeted at education, technology demonstration, or commercial use. Of the more than 1000 CubeSats launched through December 2018, 107 have been identified as scientifically motivated. Some of these have returned publication quality data, revealing aspects of the space environment not well studied by their larger brethren. Fig. 1. 5 shows the number of scientific CubeSats launched by year. Approximately half of these have only been launched since early 2017, including

36 which are part of the QB50 constellation, discussed in more detail below. 49 of the 107 scientific CubeSats have been declared successful, meaning that the primary mission objectives have been met or the satellite is taking actions that are anticipated to achieve primary mission success. However, some of the recently launched CubeSats are still in commissioning or early operations, thus their scientific productivity is yet to be ascertained. The number of "successful" missions can thus be expected to increase in time.

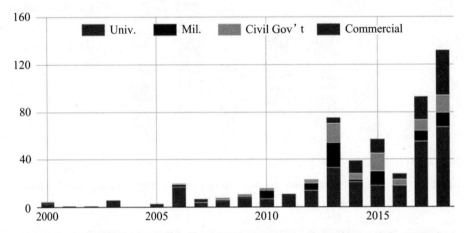

**Fig. 1.4** **Annual number of CubeSats launched by type of organization responsible for design/construction/operation. The more than 450 commercial constellation CubeSats developed by Planet and Spire beginning in 2014 are omitted from the chart for readability. Chart created by M. Swartwout using data through the end of 2018 (data from https://sites.google.com/a/slu.edu/swartwout/home/cubesat-database)**

In 2017, 36 CubeSats were launched to explore the upper atmosphere as part of the EU-organized QB50 constellation.[①] The constellation included CubeSats contributed by many countries, such as Australia, the U.S., Canada, China, South Korea, Israel, South Africa, Turkey and Ukraine. The QB50 project administration provided each group with extensive technical and administrative support, including professional design reviews, a science payload and complete launch campaign. Some teams also received a full ADCS bundle free of charge. The teams were required to invest an additional 600 – 700 k€. The QB50 project can be viewed as a sort of pathfinder for international constellation missions, in which individual countries

① https://www.qb50.eu/.

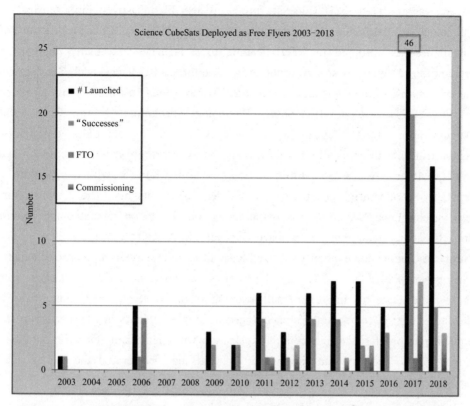

Fig. 1. 5 Scientific CubeSats: 107 launched from 2003 to 2018; 46 in 2017 alone. 7 of the 107 were lost due to launch failures (labeled FTO in the plot; FTO = Failed to Orbit). Two mission status categories not plotted are those that were Dead on Arrival (21), and those whose mission status is not known/reported (14) (data taken from Swartwout, https://sites. google. com/a/slu. edu/ swartwout/home/cubesat-database)

contribute a complete spacecraft rather than a single instrument or subsystem. This is discussed further in Section 3. 5. 1.

The majority of scientific CubeSats built in the U. S. have been supported by the National Science Foundation (NSF). Until recently, NASA CubeSats were primarily focused on technology development. After the success of the NSF CubeSat program, funding opportunities for scientific CubeSats increased at NASA, and more than twenty scientific CubeSats have launched or are in development as of this writing.

In Europe, the interest in CubeSats for science has also been increasing. Starting in 2005, the European small satellite effort initially focused on hands-on educational projects with 1U to 3U CubeSats produced by universities. In 2010, interests changed towards larger 3U to 12U CubeSats produced also by industries and agencies for technology demonstration. In Germany, a dedicated educational CubeSat program for universities was initiated in 2009, leading to 6 launches and many missions in preparation. Now, Ireland is building its first satellite, a 3U science CubeSat that will measure cosmic gamma ray bursts. [①]Since 2013, more than 10 MEuro were dedicated to ESA GSTP (General Support Technology Programme) for 7 In-Orbit Demonstration (IOD) CubeSat missions. These include several LEO constellation demonstrators with applications in $NO_2$ pollution monitoring, weather prediction and space weather. A Ka-band interferometry swarm (KRIS) will demonstrate a capability for measuring ocean currents and sea surface heights. The demonstrators also include Lunar CubeSats for mapping ice on the moon and studying meteor impacts, and a stand-alone deep space CubeSat (M-ARGO) (Walker, 2018). The European Commission explicitly referred to small satellite missions in its programmatic work program 2018-2020, "The development of new and innovative approaches, such as the use of CubeSats and other small space platforms, or the use of Commercial off-the-shelf (COTS) components is encouraged as long as it leads or contributes to the implementation of space science and exploration with significant scientific outputs." There is little question that the rise of CubeSats has accelerated the use of small satellites for science.

> **Finding 1.1** Small satellites across the full spectrum of sizes, from CubeSats to ~300 kg microsatellites, have enabled important scientific advancements across the space sciences.

> **Finding 1.2** Small satellites, particularly CubeSats, have enabled access to space for more nations, and have provided opportunities for countries with new or small space programs to participate in much larger international projects.

> **Finding 1.3** The emergence of CubeSats has resulted in a significant increase in launch cadence. However, the launch cadence of larger, traditional small satellites has decreased in the past few decades, and the development time and cost have not decreased.

---

① https://www.rte.ie/news/business/2017/0523/877210-satellite/.

# 1.1.3 Launch opportunities, commercialization, and other developments

One of the limitations in the early development of the smallest satellites was the limited availability of launch opportunities. The 1960s had the first boom of rideshares, followed by a lack of rideshares for a number of years. The number of rideshares increased substantially around 2007, primarily due to the advent of the CubeSat standard and launchers. Standardization has permitted the CubeSat-carrying canisters to be qualified for launch as hitchhikers on more than a dozen different launch vehicles, including the International Space Station. This standardization, along with relatively low cost to develop and launch CubeSats, has led to explosive growth in application areas well beyond education and training. Recently, CubeSats are being used by the commercial sector as elements of global constellations of hundreds of satellites.

The increased interest of small satellites in the commercial sector promises to generate new launch opportunities and drive down launch costs. A comprehensive analysis of market drivers and access to space is given in Chapters 3 and 4 of Lal et al. (2017), respectively. Today, relatively frequent and cheap launch opportunities into low-Earth orbit (LEO) exist. These include launch from the international space station and piggyback opportunities on PSLV (India), Dnepr and Cosmos (Russia), Long March (China), Vega (Europe), and Falcon (USA). In some cases, rockets are shared by many small satellites (such as the PSLV-launch in February 2017, which carried 104 small satellites). There are several new companies building rockets for small satellites (e. g., Rocket Labs and Virgin Orbit). Brokerage organizations offer integration of spacecraft and find launch opportunities by contacting organizations that have launch capability to the desired orbit. A more detailed discussion of launch-related policy issues is discussed in Section 3.2.3.

The commercialization of small satellites and entry by new players is also increasing the availability of commercial off-the shelf complete subsystems which has the potential to significantly reduce cost and development time for scientific satellites. Commercial parts are already being used for scientific CubeSats, and reliable mass production of parts has already been demonstrated. For example, the company ISIS

(Innovative Solutions in Space)① has supplied components and subsystems for 260 small missions. NASA's Small Spacecraft Virtual Institute② now provides a parts search tool for users to obtain information about commercial parts survivability. Such resources will help provide a justification for using such parts in scientific satellites that are bigger than a CubeSat.

The developments in the small satellite sector could create a new paradigm for small scientific satellites. The combination of low launch costs, COTS parts, and ability to purchase a complete satellite could drive down the cost of a mission to the point that it will be cost effective to streamline testing, structural verification and analysis. In addition, smaller facilities can be used for testing of small satellites thus reducing cost. The commercial sector is already developing new ways to build and test small satellites. For example, OneWeb Satellites recently set up production lines to manufacture up to three satellites per day using aircraft manufacturing technologies. ③ OneWeb has ordered 900 of the 150 kg satellites for less than a million dollars per unit (Iannotta, 2019).

> **Finding 1.4**  The rapid increase in CubeSat launch cadence can be attributed to standardization which increases rideshare opportunities, cost reduction due to availability of COTS parts, and an explosion of their use in the private sector.

> **Finding 1.5**  The cost effectiveness of increased ride-share opportunities and larger launchers, in combination with smaller spacecraft and low-cost COTS parts has already enabled large constellations, e.g. Planet and QB50, opening up new opportunities for science.

Over the past three decades, advances in technology have revolutionized the way we live, work, and drive. Yet, the technologies that have given us the internet, smart phones, and much safer and smarter cars are not in our missions flying in space. Instead, most science missions are constructed from parts that existed more than a decade ago. This is necessary for large, expensive missions which must use qualified parts with a high reliability. However, a lower cost mission can tolerate more risk,

---

① ISIS, Motorenweg 23, 2623 CR, Delft, The Netherlands, https://www.isispace.nl.

② https://www.nasa.gov/smallsat-institute.

③ http://www.oneweb.world/press-releases/2017/oneweb-satellites-breaks-ground-on-the-worlds-first-state-of-the-art-high-volume-satellite-manufacturing-facility.

taking advantage of newer technologies that haven't been space qualified to the nth degree. So far, this is only happening with CubeSats and not with the traditional, larger (~100 kg) small scientific satellites.

The result is that traditional small science satellites have not seen the reduction in cost or development time that is being achieved in the commercial sector. Since the restart of the NASA Explorers program in 1988, the average time from selection to launch of a SMEX mission has been 5.6 years. This does not include the time between the NASA Announcement of Opportunity and the selection, about 9 months, nor does it include the preparation time for development of a mission concept that is sufficiently mature to have a reasonable chance of selection, which may take a few years. As a result 7.5 ± 2 years pass between an instrument concept and the start of scientific analysis.

Small science missions take too long and are more expensive than they need to be, leading to long wait times between proposal opportunities, and low proposal success rates. This only reinforces the risk aversion that is present in the selection process and management structure of even the small missions led by national space agencies. Scientists are discouraged from innovating and taking risk because they may only get one or two chances in their entire career to lead a mission. This risk aversion potentially leads to mediocre or incremental science. The boom in commercial small satellites offers the chance to find a new way of doing business, presenting a real opportunity to change the paradigm for small satellites in space science (Section 3.3 and 3.4).

> **Finding 1.6**  The science community has not yet fully capitalized on advances in technology or the increased activity in the commercial sector in order to reduce the cost or development times of traditional small satellites. A lack of frequent flight opportunities persists, potentially discouraging innovation by sponsoring agencies and scientists.

# 1.2   Scientific potential of small satellites and CubeSats

In this section we examine the near-term scientific potential of small satellites, and highlight a few mission concepts that are currently under development. This section is not intended to be comprehensive, rather, we hope to illustrate the wide range of science applications currently employing small satellites. We also consider current limitations and challenges for using small satellites for space science.

## 1.2.1 Overview

The importance of traditional small satellites has been recognized and reaffirmed by the science community, particularly in Astrophysics, Heliophysics, and Earth Sciences. The most recent U. S. Decadal Surveys[①] in all of these disciplines recommended augmentations of NASA's traditional small satellite programs (i. e., Explorers and Earth Ventures). Even smaller satellites, such as CubeSats, are enabling new kinds of science. The 2016 US National Academies report, "Achieving Science with CubeSats" (NASEM, 2016)[②] provided a comprehensive overview of the scientific potential of CubeSats. The report concluded that CubeSats don't replace larger missions, rather they can be used to achieve targeted science goals and can also enhance larger missions by providing supporting measurements. In solar and space physics, the report found that CubeSats can provide novel measurements, for example from high risk orbits, augment large facilities, and have the potential to enable constellation missions. Constellation missions have important applications in Earth Sciences as well. Because of their shorter development time, CubeSats also have the ability to mitigate gaps in long-term Earth monitoring and are potentially more responsive to new observational needs. In Astrophysics, the small size of CubeSats limits the aperture and thus the types of science that can be done with a single CubeSat. However, the report highlighted a few capabilities, including CubeSats that stare at a single object for long periods for both exoplanet studies and stellar variability studies. CubeSat constellations may also pave the way towards space interferometers. In planetary sciences, CubeSats can provide unique vantage points and explore high-risk regions, perhaps in tandem with a larger "mother ship". They also serve as microgravity laboratories. The report noted that CubeSats have already delivered high-impact science in some of these targeted areas.

The rapid development of CubeSat technologies has already enabled new kinds of SmallSat missions. The CYGNSS mission provides a good example. The eight 28 kg

---

① Decadal Surveys, published by the U. S. National Academies, can be found at http://sites. nationalacademies. org/ssb/ssb_052297.

② Available at https://www. nap. edu/catalog/23503/achieving-science-with-cubesats-thinking-inside-the-box.

spacecraft are not CubeSats, but used commercial parts and a tailored mission assurance approach. The resulting cost was $5 M/spacecraft (not including payload), compared to $165 M/spacecraft for the MMS mission (Tumlinson, 2014). A precursor of CYGNSS was TechDemoSat-1, weighing in at 157 kg (Foti et al., 2017). It was launched in 2014 to demonstrate the method, also used by CYGNSS, of using Global Navigation Satellite Systems-Reflectometry (GNSS-R) for observing hurricanes. While TechDemoSat-1 already fit the envelope of a small satellite, the CubeSat developments leveraged by CYGNSS allowed for the factor of 5 decrease in mass, thus enabling a small constellation of satellites that can monitor the development of a hurricane on relevant timescales.

## 1.2.2　Near-term science potential: Missions on the horizon

A number of recent missions and missions under development utilize small satellites. SmallSat missions don't have to be small missions; missions using a number of distributed small satellites have been launched recently and more are on the horizon.

In the U.S., the newest Explorers, TESS (362 kg, launched in April 2018 to search for nearby extrasolar planets) and ICON (291 kg, expected to launch in 2019 to study the ionosphere) are both MIDEX missions, each costing in excess of $200 M. The last SMEX missions were launched in 2012 (NuSTAR) and 2013 (IRIS). However, in response to recommendations in both the Astrophysics and Heliophysics Decadal Surveys, NASA has recently increased the cadence of SMEX and MIDEX opportunities. In Astrophysics, the Imaging X-Ray Polarimetry Explorer (IXPE) will launch in 2020 to study X-ray production in compact objects such as neutron stars and black holes. Five Heliophysics SMEX concepts were selected for further study in 2017 along with several missions of opportunity, some of which employ CubeSats.[1] A downselection to one or two missions is expected within the next year.

The number of NASA-funded scientific CubeSat launches has increased significantly in recent years (2 launched in 2015 versus 8 launched in 2018 and at least

---

[1]　https://www.nasa.gov/press-release/nasa-selects-proposals-to-study-sun-space-environment.

9 more planned for 2019). ① Among those recently selected, GTOSat, will be the first scientific CubeSat to operate in geostationary transfer orbit. It will provide key observations of the radiation belts and, with its radiation-hardened 6U bus, could serve as a pathfinder for future magnetospheric constellation missions. ② In Earth Sciences, the currently flying CYGNSS and future TRO-PICS missions are demonstrating the utility of a constellation approach to Earth Science. In particular, TROPICS, consisting of 12 CubeSats, will provide 30-minute revisit rates critical for monitoring rapidly developing storm systems (Fig. 1. 6). Future LEO constellations could exploit GPS-based relative positioning techniques for precise autonomous determination of the relative positions of the formation members, which is required for formation acquisition and maintenance, and scientific objective achievement (Causa et al. , 2018).

In Europe, traditional small satellites continue to be used for science at a relatively low rate but pursuing important science goals. The CNES MICROSCOPE mission (330 kg) was launched in 2016 to test the Equivalence Principle to one part in $10^{15}$, 100 times more precise than can be achieved on Earth (Touboul et al. , 2017). PROBA-3, one in the series of ESA PROBA missions, is expected to launch in 2020. It consists of two spacecraft with masses 340 kg and 200 kg, flying 150 m apart to create an artificial solar eclipse, allowing for study of the solar corona. The mission will also demonstrate precision formation-flying. ③ The first ESA S-class mission, CHEOPS ($\sim 300$ kg) will launch in 2019 to characterize known exoplanets. In particular, CHEOPS will measure planetary radii, which combined with the mass as measured from the ground, will allow for determination of the exoplanet density for the first time.

The number of CubeSat missions on the horizon seems to be growing in Europe.

---

① NASA science CubeSat missions were identified using October 2017 presentation by Larry Kepko, "SMD CubeSat Program Update" retrieved from https://smd-prod. s3. amazonaws. com/science-red/s3fs-public/ atoms/files/Kepko-SmallSats-APAC_October％202017. pdf on February 20, 2019. Launch dates were taken from CSLI website: https:// www. nasa. gov/content/past-elana-cubesat-launches except for MarCO launch date which was taken from the MarCO mission website: https:// www. jpl. nasa. gov/cubesat/missions/marco. php.

② https://www. nasa. gov/feature/goddard/2018/nasa-s-new-dellingr-spacecraft-baselined-for-pathfinding-cubesat-mission-to-van-allen-belts.

③ http://www. esa. int/Our_Activities/Space_Engineering_Technology/ Proba_Missions/ About_Proba-3. Retrieved July 8, 2018.

Fig. 1.6    Mission overview of the proposed NASA TROPICS mission. Image Credit:
TROPICS Team, MIT Lincoln Laboratory (The Time-Resolved Observations
of Precipitation structure and storm Intensity with a Constellation of
SmallSats (TROPICS), MIT Lincloln Lab, https://tropics. ll. mit. edu/
CMS/tropics/Mission-Overview)

ESA foresees a 8 MEuro GSTP budget for one or more projects with 3 year duration,
targeting significant improvements in system performance or new applications. An
example CubeSat mission is HERMES (High Energy Rapid Modular Ensemble of
Satellites), [1] an Italian mission for high energy (keV-MeV) astrophysics, a science
domain previously limited to large space missions. HERMES consists of a constellation
of nanosatellites (<10 kg) in low Earth orbit, equipped with X-ray detectors with at
least 50 cm² of active collecting area between a few keV and ~1 MeV, and very high
time resolution (ls). The main science goal is to study and accurately localize high
energy astrophysical phenomena such as Gamma-Ray Bursts, electromagnetic
counterparts of gravitational waves (caused by coalescence phenomena of compact
objects, such as those recently observed by the Advanced LIGO-Virgo observatories),
and high-energy counterparts of Fast Radio Bursts. A technology pathfinder consisting
of three units is under development, with a launch goal around 2020, to be followed by
a scientific pathfinder mission. The final goal is a constellation of tens of units on

---

[1]    http://hermes. dsf. unica. it/index. html.

different orbits, to provide transients positions with accuracy better than 1 degree over the full sky.

A number of international collaborative small satellite missions are currently under development. The Israeli and French space agencies (ISA/CNES), have jointly built and launched VEN$\mu$S (Vegetation and Environment Monitoring on a New Micro-Satellite). This 250 kg (dry mass) small satellite, is equipped with a super spectral camera that observes in 12 wavelengths simultaneously. The satellite provides frequent revisits (up to two days) of scientific sites spread worldwide to study the evolution of vegetation, and also serves as an in-flight qualification of a unique electrical propulsion system based on Hall-Effect thrusters. Such a system allows for minimizing the mass of propellant and utilization of non-toxic xenon, while achieving flexible orbital maneuvers. The SHALOM Mission is a joint initiative of ISA and ASI to develop several small satellites in the fields of communication and earth observation that enable the discovery and identification of contaminants on the earth's surface, in bodies of water, and in the atmosphere.

Low frequency radio space interferometer concepts are currently being explored by several nations (Fig.1.7). The OLFAR (Orbital Low Frequency ARray) concept comprises a large constellation of small spacecraft in orbit around the Moon (Rotteveel et al., 2017). The first step towards realizing this mission was recently taken with the launch of the Netherlands Chinese Low Frequency Explorer (NCLE) on 21 May 2018 (Castelvecchi, 2018).

In the future, an interferometer could even be used at infrared or optical wavelengths. A collaboration between CalTech, Jet Propulsion Laboratory (JPL), University of Surrey, and the Indian Institute of Space Science and Technology (IIST) is developing the AAReST (Autonomous Assembly of a Reconfigurable Space Telescope) mission concept to produce an optical telescope with "primary mirror" made up of distributed 10 cm-diameter circular mirrors attached to a cluster of CubeSats (Sweeting, 2018).

In recent years, microsatellites and nanosatellites have also started venturing into deepspace, beyond low Earth orbit, taking advantage of ride-share opportunities. These missions so far try to answer focused science investigations or test new technologies, in contrast with typical deep-space missions which use high-TRL parts and carry a suite of instruments.

PROCYON (Proximate Object Close Flyby with Optical Navigation), the first microsat deep-space mission (67 kg launch mass), and the first deep-space mission by a university, was launched in 2014 as piggyback of Hayabusa 2. It escaped the Earth

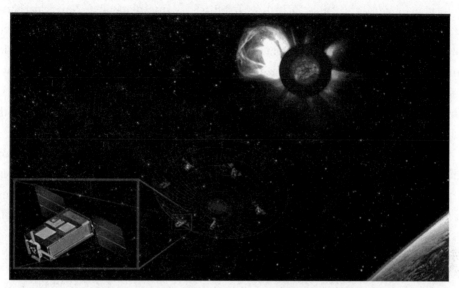

Fig. 1. 7　A number of space interferometer concepts are currently being explored including the NASA SunRISE mission concept shown here which would use a small constellation of 6U CubeSats to measure solar radio bursts. Image Credit: SunRISE team

gravity and returned one year later for a distant flyby. PROCYON validated a fully capable bus, with low, middle and HG antennas, reaction wheels and cold gas jets, electric propulsion systems, telescope and cameras. PROCYON was proposed, developed, and launched in just about 14 months, and most of the mission team consisted of students of the University of Tokyo. Two Japanese follow-on missions, EQUULEUS and OMOTENASHI, are both 6U CubeSats being developed by JAXA and University of Tokyo. EQUULEUS will use water resistojet thrusters to be the first CubeSat to go to the Lunar Lagrange point. OMOTENASHI will be the smallest Lunar lander. A demonstration mission, EGG, was recently deployed from the ISS to test a deployable aeroshell that might be used in the future for atmospheric entry or orbital insertion.

The two Mars Cube One (MarCO) CubeSats recently completed their mission to Mars, where they provided data-relay capabilities for the Entry, Descent and Landing operations of the InSight lander. ESA's Hera mission (previously AIM/AIDA) will carry two CubeSats stowed in a mothership which will deploy close to the target

Didymoon (Perez et al., 2018).① INSPIRE (Interplanetary Nano-Spacecraft Pathfinder in Relevant Environment) (Klesh et al., 2013) and DISCUS (Deep Interior Scanning CubeSat) (Bambach et al., 2018) are further demonstration projects with the objective to open deep space to CubeSats. The two spacecraft will carry a science vector magnetometer and an imager. Thirteen more CubeSats are almost ready to fly to the Moon and beyond thanks to the Exploration Mission 1, the maiden flight of SLS.②

In the future, a range of spacecraft will be available to serve a palette of mission types. These could range from high tech chipsats (attoor femtosats) to traditional large spacecraft with augmented capability based on miniaturized space technology. Large spacecraft with piggyback small satellite probes may operate within the solar system. The probes may be a part of the primary mission or they may be on-board as a result of a rideshare.

## 1.2.3 Limitations and technological challenges

The recent US National Academies report (NASEM, 2016) on CubeSats provides an overview of technologies needed for scientific advancement, along with recent technology developments. In particular, advances in propulsion, communications, sensor miniaturization, radiation tolerant parts, and sub-arcsecond attitude control were called out, among others. The IDA report on small satellites (Lal et al., 2017) provides a more recent assessment of technology trends for small satellites in general, ranging from high bandwidth communications and onboard processing to advances in miniaturization to orbital debris surveillance (Lupo et al., 2018; Santoni et al., 2018) and removal technologies. Many of the technology developments driven by commercial markets are also needed for science missions. Current technology innovation trends addressing some of the limitations of small satellites in low earth orbit include:

- *Noise reduction in miniaturized components*: Software approaches based on filter technologies reduce the susceptibility to noise.

- *Attitude and orbit control capabilities*: recent miniature reaction wheel developments improve attitude control at low power consumption and electric

---

① https://phys.org/news/2019-01-cubesats-hera-mission-asteroid.html.

② NASA's new Space Launch System (SLS).

propulsion systems provide orbit control. Thus, even for a 1U-CubeSat, improved instrument pointing and formation capabilities are being realized (e. g. , OCSD,[1] UWE-4 and TOM missions). A fully magnetic attitude control subsystem is presented by Colagrossi and Lavagna (2018).

- *Communication link capacity*: new developments on optical links promise capacities beyond 100 MB/s at clear sky (e. g. , OCSD, QUBE and TOM missions), but also very miniature X-band transceivers are becoming available (e. g. , MarCO mission).

- *Extending the lifetime*: advanced FDIR (fault detection, identification and recovery) methods and redundancy concepts guarantee reasonable lifetime in orbit, even for commercial off the shelf components (e. g. , UWE-3 has operated without any interruption for more than 3 years despite encountered SEU and latch-ups).

- *Ground segment*: several university ground station networks have been initiated (e. g. , GENSO, UNISEC) to support frequent transmission of data from small satellites in order to relieve the on-board data storage and processing requirements. Commercial networks (e. g. , KSAT lite) are providing global coverage for SmallSats.

The data return of scientific small satellites thus far, particularly CubeSats, has been limited by availability (and cost) of ground stations. Communications may become even more difficult with the development of large constellations for both commercial and scientific use. Constellations of hundreds or thousands of satellites, especially with imaging capabilities for Earth Sciences, will produce massive amounts of data. Sweeting (2018) states: "Over the next decade, the amount of data that will be cumulatively downlinked by small satellites is expected to reach 3. 9 exabytes (exabyte = 1012 MB). Traditional RF capabilities are unlikely to be able to meet this demand…". Several efforts are underway to develop low power optical terminals in space capable of transmitting data at rates up to 10 Gb/s (Sweeting, 2018). Nevertheless, on board processing in order to limit the amount of data transmitted to the ground may be required. Advances in data processing (e. g., artificial intelligence) may prove useful and necessary for the science missions of the future. Commercial constellations and operational systems are also likely to require data distribution systems that are useable in close to real time. Such systems may provide

---

[1] https://aerospace. org/story/communicating-and-converging-cubesats. Retrieved February 22, 2019.

new opportunities for science missions.

> **Finding 1.7** Technologies further enabling formation flying, intersatellite communication, data-compression and mega-constellation deployment will be in demand as scientific ambitions increase.

Unique challenges exist for deep space missions. We address these in more detail since they aren't as well covered in the reports referenced above.

*Telecommunications*: Currently, deep-space Cubesats are designed for low-data volume measurements which limits scientific observation. Moreover, deep-space missions need X-band or Ka-band on-ground antennas, and therefore the support of large space agencies with their ground station networks. In the past, ground station time has been negotiated for small missions (on an opportunistic basis and with low priority to other missions), but typically for one or two spacecraft at a time. The democratization of deep-space exploration could require supporting dozens or hundreds of nanosats, especially during the launch and early operation phase. For example, most of the 13 Cube Sats launched with EM-1 will have to perform a critical maneuver within two days of deployment, for which they need downlink and uplink for operations and for precise orbit determination (two-way Doppler, DDOR). Current developments in optical link equipment and X-band transceivers open new perspectives for solutions. An increasing number of worldwide distributed, smaller ground stations will provide continuous coverage in the future, similar to the radio amateur supported UHF ground station networks of the CubeSat community in UHF/VHF.

*Power generation*: At significant distances from the sun, energy generation requires large solar arrays (as for ROSETTA) or use of alternative energy sources. Nuclear generators (as for Cassini, or Galileo) have flown on interplanetary spacecraft but are currently not available for CubeSats. Thus, only the very limited storage resources of batteries can be used, demanding very careful operations in order to not waste those scarce resources.

*Propulsion*: Most CubeSat propulsion systems to date have been cold or warm gas systems due, in part, to their relatively low cost and low level of complexity. A broad array of various types of electric propulsion systems for CubeSats and microsatellites are in development by multiple companies at Technical Readiness Levels (TRL) between 5 and 7. [1]

---

① https://sst-soa.arc.nasa.gov/04-propulsion.

In LEO, one CubeSat mission, AeroCube 8 has already demonstrated miniature electric propulsion system capabilities. [1] For deep-space missions, (total impulse/volume) must be increased, complemented by larger fuel storage capacities. Innovative technologies like solar sailing are being considered and may provide solutions.

*Mission design*: Orbital mechanics and navigation is especially challenging for small deep-space spacecraft, which have limited orbit control capabilities, yet need to reach similar destinations as larger-class spacecraft. Mission design is as critical and complex for small satellites, as it is for large satellites, and sometimes even more so, relying on expert manpower and advanced tools. For this reason, mission design activities for SmallSats are mostly carried out by space agencies. Support toward the development of open-source (and ITAR-free) mission design tools would reduce the costs of deep-space nanosats and enable the participation of new stakeholders.

*Operations*: Like mission design, operations for a small satellite can be as complex and expensive as for a large mission. Operations of deep-space missions are mostly carried out with a man-in-the-loop approach. Autonomy would reduce mission costs, but it is currently not implemented to its full potential on expensive missions because of the associated risk. Deep space missions will benefit from automation efforts in the near-Earth environment currently being developed for swarms and formations. Deep-space nanosats, however, must rely on an even higher degree of autonomy because of the limited ground station availability. Support towards the development of autonomous operation and navigation technologies would enable deep-space exploration by small satellites, and eventually reduce the cost of large-class missions as well.

*Launch Opportunities*: Increased access to space is also needed to increase the cadence of science flight opportunities. As discussed in Section 1. 1. 3, launch opportunities have improved significantly for near-Earth missions. The promise of small satellite launchers currently being developed (Table 4. 1 in IDA report, Lal et al. , 2017) will enable CubeSats to go to a larger range of orbits, without the restriction of going where the bus is going. For example, Rocket Labs' Electron just launched 13 CubeSats into LEO in December. [2] However, for deep-space missions,

---

[1]　http://spl. mit. edu/news/aerocube-8-cd-launch-mit-spls-electrospray-propulsion-space.

[2]　https://www. rocketlabusa. com/news/updates/rocket-lab-successfully-launches-nasa-cubesats-to-orbit-on-first-ever-venture-class-launch-ser-vices-mission/ Retrieved February 22, 2019.

rideshares are more limited (Fig. 1. 8). In 2015, NOAA's DSCOVR satellite was launched on a Falcon-9 to the Earth-moon L1 point with unused capacity of 2500 kg, and NASA's TESS mission was launched into a translunar injection orbit with 3000 kg of lift capacity to spare. [1] The opportunity to routinely open up launcher capacity on lunar, lagrange point or interplanetary missions for small spacecraft would be a game changer for deep space CubeSats. Such opportunities may soon become reality: NASA recently committed to flying an ESPA ring with every science mission in order to make the excess capacity available to small spacecraft. [2]

> **Finding 1. 8**  Significant technology advancements are opening up new opportunities for small satellites, addressing challenges that have so far limited their science return. However, there are particular additional challenges associated with deep-space exploration.

In summary, the emergence of CubeSats is driving important advances in technology that are already enabling non-CubeSat small satellite missions, such as CYGNSS. While the scientific promise of CubeSat missions is high, the number of missions launched with the intent to conduct scientific investigations is less than 12% of the total number of CubeSats launched through 2017. The number of missions using larger small satellites is even smaller. Nevertheless, CubeSats or CubeSat-enabled small satellites are already being flown in large commercial constellations. The scientific potential for constellations consisting of dozens, or more, nanosatellites for space science has yet to be borne out. However, the potential for purposefully designed scientific investigations comprised of CubeSats, larger small satellites, or both working together synergistically holds great scientific promise.

# 2　Visions for the future

In this section we turn the attention from our neighborhood to the peaks in the distance, which represent visions for the future: developments or missions that are out of reach with current or imminent technology, in some cases for several decades.

---

[1]　https://spacenews. com/government-agencies-prepare-for-piggyback-flights-secondary-payloads/ Retrieved February 22, 2019.

[2]　https://spacenews. com/nasa-bolsters-smallsat-science-programs/Retrieved June 8, 2019.

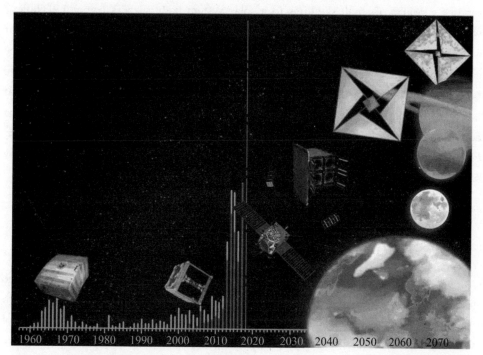

Fig. 1.8    The light blue bar graph shows the evolution from 1957 to present (based on data from Swartwout). The CubeSat numbers are indicated with dark blue. With the MarCO mission that just completed its mission to Mars, the upcoming SLS-EM-1 and the Hera mission to Didymos/Didymoon, rideshares extend beyond Low Earth Orbit. All three missions have adapted the CubeSat format. Future missions may require lower overall spacecraft mass and hence new spacecraft concepts. For instance, CubeSat swarms deployed from deep space pods or laser propelled ultra-small chip-sats. Artist: Katrine Grønlund (Image of Oscar-1 used with permission: http://www. arrl. org/space-communication; DTUsat-1 image courtesy the DTUsat-1 team; Image of Hera mission: https://www.esa. int/Our_ Activities/Space_ Engineering_ Technology/Asteroid_ Impact_ Mission/Asteroid_ Impact_Deflection_ Assessment_mission; Image of Isispace QuadPack CubeSat deployer used with permission: https://www. isispace. nl/product/quadpack-cubesat-deployer/) (For interpretation of the references to color in this figure legend, the reader is referred to the web version of this article)

Section 2. 1 deals with Earth and Geospace science, where a global system of hundreds or even thousands of small satellites, all communicating with each other and with the ground, will have enormous impact not only on science, but also on society (although the latter is outside the scope of this document). Section 2.2 explores the potential of sending a swarm of small satellites to a solar system body such as comet 1P/Halley, when it returns to the inner solar system in 2061. Section 2.3 describes the potential of a synthetic aperture optical telescope made up of small satellites, which might be capable of imaging stars other than the Sun. Finally, in Section 2.4, we consider the possibility of an interstellar mission based on the Breakthrough Starshot initiative that has attracted much public attention recently. These four visions are not meant to be comprehensive or authoritative in any way, but rather serve as examples of what might become possible when projecting the potential of small satellites into the future by several decades.

# 2.1　Potential of small satellites for Earth and Geospace sciences

CubeSats and SmallSats have the potential to make unique contributions in a range of science domains within Earth observation and Solar and Space Physics. In particular, SmallSats will enable large constellation missions, thus providing a new tool for doing science from space. In Earth Science, applications include surface imagery, meteorology, studying pollution, and measurements of the solar irradiance, to name just a few examples. In space physics, the importance of studying the Sun-Earth system using a systems-science approach was highlighted in the last US Decadal Survey. The recent National Academies report on CubeSats (NASEM, 2016) emphasized the importance of multipoint measurements to accomplish this and hence a major advantage of SmallSats.

This section briefly describes a few notional missions that are nearly within reach due to opportunities created by SmallSats. The idea for such mission concepts is not new (e.g., Esper et al., 2003); the challenge for this vision is less about needed technologies, and more about feasibility within available budgets. However, developments in the commercial sector (Fig. 2.1) may provide pathways to reducing cost and achieving such visions.

## 2.1.1　Mega-constellation for Earth Science

There are countless applications in Earth Science for a large constellation (i.e., hundreds to thousands) of LEO satellites. Smaller constellations consisting of about 5~10 small satellites are already flying, such as CYGNSS (Ruf et al., 2018),

Fig. 2. 1   **Schematic of a large commercial satellite constellation. Image Credit: Telesat**

discussed in Chapter 1. Specific mission profiles will differ in terms of spatial, temporal, and spectral resolution depending on the application, but, generally speaking, the use of SmallSats (similar to large EO satellites) to monitor the Earth should strive towards providing as much data, as often, as accurate, as precise, and as complete (wavelength, polarization) as possible.

Using small satellites for monitoring the Earth has the following main advantages:

• Using a large number of these satellites in a constellation increases the revisit frequency, which allows for studying changes over short time intervals, or for tracking purposes.

• The production of a large number of these satellites may enable cost reduction based on standardization and miniaturization.

• As launch costs are proportional to the mass, it is cheaper to launch small satellites, even to build a constellation, compared with the cost of a single large satellite.

• Small satellites can be used for demonstrating the feasibility of new mission concepts for larger missions.

However, these advantages should be traded against the laws of physics (e. g., resolution$\simeq \lambda/D$) which still need to be obeyed. Technology used for large satellites needs to be adapted to the constraints set by small satellites, mainly size and mass. In order to compensate for the small size of a small satellite, spatial resolution can be improved by using higher operating frequencies or by artificially increasing the aperture D. For example, techniques such as synthetic aperture radar and interferometry could be feasible with constellations of small satellites. Both are associated with technical challenges (e. g., high-frequency receiver/transmitter and control/stability, propulsion) but one should not exclude such promising developments in the future.

Another important potential for SmallSats is working in symbiosis with large satellites in order to complement their capabilities. Such an approach is already used by Landsat or by the Copernicus/Sentinel-2 system, combined with Planet data, for which accurate multispectral data are complemented by daily high-resolution images. Another example is the ESA Earth Explorer mission called Fluorescent Explorer (FLEX), which is being designed to fly with Copernicus Sentinel-3. This synergy of missions can also be used to develop techniques for ensuring the cross-calibration of measurements between different missions and ensuring the quality of these data for science applications.

Innovative concepts for small satellites, e. g. passive receiving only radar antennae flying together with other, larger satellites acting as the transmitters for bistatic measurements, are under development. The recent ESA definition phase of the Lband SAOCOM Companion Satellite (CS) demonstrated that a passive small SAR could be developed to significantly improve the science mission objectives of the main mission (flying with the larger SAOCOM mission). The agility and some of the techniques associated with small satellites, if properly mastered, could open (as in some cases already happening today) many other new applications linked for instance to video capabilities, real-time imaging, and instruments directly commanded by the users on the ground.

Worth noting is that one should not only consider high revisit rate vs. (spatial) resolution, concluding that the former is more important to customers than the latter (which is usually, but not always, true). In fact, spectral and/or radiometric parameters can also be "key application enablers" for atmosphere or hydrosphere observations and even for "classical" land imagery. Furthermore, actual mechanisms and feedback loops can be captured by multipoint measurements, perhaps even using different techniques such as optical and SAR. This would allow for capturing interactions between the various cycles, which would be a paradigm shift rather than an incremental improvement in spatial or temporal resolution.

In order to realize such a mission, opportunities exist for science to take advantage of developments in the commercial sector. As discussed in Chapter 1, commercial interest in small satellites for Earth Observation is growing at a rapid rate. For example, in 2017, 328 small satellites were launched, 103 of which were in a single launch (Planet with ISRO/PSLV launched on 14 Feb. 2017). Out of the 328 satellites, two thirds were used for Earth observation, with masses from <10 kg (89%), 10~100 kg (7%), up to 100~500 kg (4%); the remaining one third was for technological and scientific applications (31%) and for communications (2%) (all numbers from CATAPULT, 2017). It is worth noting that the total mass of these 328 satellites was less than the 8-ton ENVISAT launched by ESA in 2002 carrying ten different instruments. [1]

Scientific constellation missions could leverage the technology (standardization and miniaturization), and learn from industrial manufacturing and I&T methodologies. Finally, opportunities may exist for commercial data buys or for putting science instruments on commercial platforms. So far, such partnerships are challenging to develop, though successful examples do exist.

Regardless of the model employed, some key elements of the data acquired by small satellites should be considered including:

• *Sustainability of the data*: How to ensure, from a science perspective, that the data acquired by these missions are based on a long-term commitment needed by most science users.

• *Data policy*: How to guarantee a data policy which is as open, full and free as possible for all the data, which is a precondition for a good science development plan[2].

• *Potential conflict of acquisitions between commercial and science requirements*: Many recent developments of SmallSats are driven by commercial entities which make available large datasets of EO data. When these data are used for scientific purposes, it's important to ensure that commercial interests do not jeopardize the science potential of such missions.

---

[1]  Interestingly, the cost per kg per instrument for ENVISAT yielded approximately 0. 3 M$ /kg, which was very competitive compared to present-day SmallSats with typically single instrument payloads launched in 2017, whilst also considering that ENVISAT lasted for ten years from 2002 to 2012 (twice its nominal design lifetime).

[2]  https://earth. esa. int/web/guest/-/revised-esa-earth-observation-data-policy-7098, 2010, or European Commission Delegated Regulation No 1159/2013, https://eur-lex. europa. eu/legal-content/EN/TXT/HTML/? uri=CELEX:32013R1159&from=EN.

• *Compatibility*: SmallSats should be considered an element within a bigger ecosystem, often flown to add additional capability to the institutional satellites such as the Copernicus Sentinels.

• *Data downlink*: Large amounts of data may require on-board data processing/handling to downlink only useful data (e. g. , cloudy images to be disregarded by on-board processing).

• *Data usage*: Once the data are on the ground, the forth-coming challenges for SmallSats are more in the exploitation of large data sets (big data, artificial intelligence, merging different types of data-satellite and non-satellite)-rather than in the development of the satellites themselves.

More generally speaking, the development of SmallSats for science should answer the need of the users' community in order to avoid a technology push approach, which might generate a deluge of uncalibrated and useless data. This would also avoid over-selling of SmallSats, i. e. making promises that cannot be kept, which might be detrimental to the further use of this type of satellite by the science community. The importance of open data policies and international sharing of data cannot be overstated. The Move-Bank initiative, a database for collecting and distributing data of migrating animals collected by biologists worldwide, may be used as a model or inspiration for such a data share initiative. ①

## 2.1.2 Magnetospheric constellation mission

The important role of small satellites for magnetospheric research is well accepted (e. g. , Shawhan, 1990), and the space physics community has discussed the need for a large magnetospheric constellation mission for decades (e. g. , Angelopoulos and Panetta, 1998; Fennell et al. , 2000). Smaller constellations of 3-5 spacecraft, such as Cluster, THEMIS, SWARM, and MMS, have already enabled transformational science. THEMIS serves as a particularly good pathfinder for using SmallSats; its five 77 kg (dry mass) satellites, launched on a single Delta II rocket, have been operational for almost 12 years. However, the leap from 5 satellites to dozens or hundreds of scientific satellites has not yet occurred.

The magnetosphere, ionosphere, and thermosphere together act as a coupled system, responding to driving by the solar wind from above and the lower atmosphere

---

① https://www. movebank. org/.

from below. Understanding how the energy of the solar wind couples into this system and the interaction between adjacent regions of space requires multipoint measurements over broad regions. Like weather stations distributed around the globe or buoys across the ocean, distributed measurements throughout the magnetospheric system, coupled with sophisticated computational models (Fig. 2.2) (Spence et al., 2004), will transform our ability to understand and make predictions about the space environment.

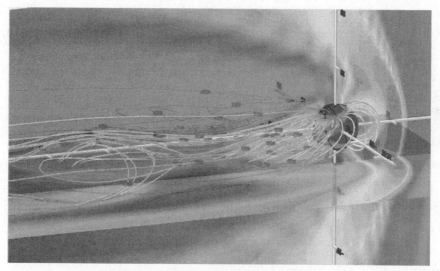

Fig. 2. 2　The Magnetospheric Constellation (MagCon) mission concept from a NASA mission definition study showing 36 spacecrafts superimposed on a magnetohydrodynamic (MHD) simulation of Earth's magnetosphere. Figure reproduced from Spence et al. (2004)

Such a mission concept has so far continued to be decades away. However, the rapid pace of development of small satellites gives reason to be optimistic. Efforts are underway to develop minitiarized instrumentation. For example, the 6U Dellingr CubeSat, developed at NASA's Goddard Space Flight Center, provided a test bed for a small science magnetometer and mass spectrometer. The CSSWE CubeSat developed at the University of Colorado used a miniaturized version of the Van Allen Probes REPT energetic particle instrument. The Goddard team is currently working on a radiation hardened bus (GTOSat) based on the Dellingr design. The next critical step will be learning how to manufacture and test large numbers of identical satellites, an area about which the science community must learn from industry.

It may also be possible to leverage commercial space to realize some components of a large constellation. For example, hosted payloads on commercial LEO satellites could target specific measurements at the ionospheric boundary. An already existing example is the AMPERE project, funded by the US National Science Foundation, which uses the ADCS magnetometers on the Iridium satellites to detect field aligned currents in the auroral region (Anderson et al., 2000). The project was achieved with a public private partnership between the commercial space industry, university researchers, and NSF (government). A more recent study used magnetometers on the Planet Labs Inc. CubeSat constellation (Parham et al., 2019). Providing a different view, the recently launched GOLD mission makes measurements of the upper atmosphere from geosynchronous orbit. GOLD uses a science instrument—a UV spectrograph—that is hosted on SES-14, a commercial communications satellite built by Airbus for SES Government Solutions.

International collaboration is another means to achieve an ambitious vision such as a magnetospheric constellation. In fact, the QB50 project discussed in Section 1.1 included plasma instruments (Langmuir probes) on some of the spacecraft. Although the project had goals other than scientific research, it can serve as a model of international collaboration that enabled a large number of small satellites to be built and launched in a coordinated way (see also Section 3.5).

## 2.1.3　Conclusions and findings

SmallSats enable new science, applications and commercial developments at all levels (upstream, downstream, national, international) especially via constellations and convoys of multiple satellites combined with readily available platforms and short R&D update cycles. However, development in Earth Sciences cannot be done cheaper, faster, better using SmallSats alone, and these SmallSats should be considered elements of a larger measurement ecosystem.

Nevertheless, this new generation of satellites offers opportunities worth exploring and developing to support a better understanding of Earth as a system, including addressing observational gaps and providing more frequent measurements. In order to be beneficial for the science community, one should ensure that key issues linked to a free, full, and open data policy, generation of useful and well-calibrated data, and ensuring a long-term and sustainable stream of data, are taken into account when considering the development of this promising new domain.

Finding 2.1 An opportunity exists for transformational advancements in Earth and space sciences using large constellations of satellites. This vision may be achieved with stand-alone science missions or through partnerships with industry that make use of the increasing number of small satellites in orbit. The scientific community would benefit tremendously from the data acquired by this large number of satellites assuming these are governed by a free, full and open data policy for research purposes.

## 2.2 Swarm exploration of a solar system body

This section elaborates on one high-impact planetary science concept, then provides a couple of additional examples of science applications that would benefit from large constellations (networked or not) of CubeSats or SmallSats.

### 2.2.1 Exploration of "Once in a Lifetime" planetary bodies

This concept targets planetary objects with very long periods (referred to as LPOs), i.e., bodies that cross our solar system and approach Earth only once in a person's lifetime. These bodies include Oort cloud comets (200 + years), Manx objects and, now, interstellar objects (ISOs), such as the recently discovered 'Oumuamua (Meech et al., 2017). This is not the first, and certainly not the last interstellar visitor in our solar system. Long-period comets are the most primitive witnesses of the early solar system. Interstellar visitors are suggested to be ejecta of extrasolar planets subject to catastrophic collisions. Hence the scientific value of exploring these objects is unbounded, especially as a recent study suggested that these collisions could have offered a means to transfer life organisms among extrasolar systems (Berera, 2017). This discovery carries implications on a fundamental level regarding the place of humanity in the universe and the prospect to sample extrasolar planets.

A very broad range of measurements are sought for long-period and Manx comets and ISOs. They include basic physical properties characterization (shape, density, morphology, dynamical properties), compositional properties (elemental composition, mineralogy, isotopes of at least hydrogen, oxygen, nitrogen, and carbon), geophysical/interior properties (porosity, cohesion, magnetic field), geological traits that might inform on origin and possible long-term evolution, and interactions, in particular of a coma when it exists, with the solar wind. Instruments small enough to perform these measurements already exist but their operation might prove challenging, as described in more detail below. Instruments of choice include dust spectrometers (mineralogy, dust coma density) because they can operate when interacting with high-velocity material; in situ remote sensing instruments such as submillimeter wave spectroscopy (e. g., the MIRO instrument on Rosetta), which allows constraints on isotopic properties of volatiles from a safe distance; other in situ remote sensing instruments include color imagers, and spectrometers covering a broad range of wavelengths. Elemental measurements are more complex to implement in that they require close interaction with the target for some extended collection time. Elemental abundances may be obtained in part from measuring the plasma generated between the target's coma and/or dust and the solar wind.

The exploration of LPOs is challenging for many reasons: ① the orbital properties of these bodies are not known with enough lead time to develop a mission; ② they have a broad range of inclinations; ③ the encounter velocities are in excess of 50 km/s, hence the encounters may be very short; ④ LPOs may be geophysically active or made of multiple coorbiting elements. The only attempt to explore a comet with a longer period (~75 years) up-close was the encounter with comet Halley in 1986. Its visit was deemed such an important event that six spacecraft were sent by different space agencies: ICE (NASA), VEGA 1 and 2 (Roskosmos), Suisei and Sakigake (ISAS, its first science mission), and Giotto (ESA). The deployment of three spacecraft at once was and remains the first instance of its kind. The missions were coordinated by the IACG (Inter-Agency Consultative Group), which was created for this purpose. NASA's ICE did not in fact "encounter" Halley as it stayed outside the shock front, yet it was important to help showing the others the way.

The challenges in implementing a mission to an LPO or Halley during its 2061 return may be addressed by sending a very large number of spacecraft separately by multiple space agencies, and in a coordinated manner (Fig. 2.3). It is simply too big an endeavor to expect any single space agency to send a very large number of assets with a diversity of capabilities commensurate with the broad science knowledge sought at these bodies, within today's budgets. On the other hand, the enormous interest generated by the visits of LPOs and 'Oumuamua on a worldwide scale indicates that an international effort to coordinate future exploration of these bodies is a worthy and realistic endeavor.

Constellations, formations, and swarms of small spacecraft have been identified as game changers for enabling new space science (NASEM, 2016). In recent times,

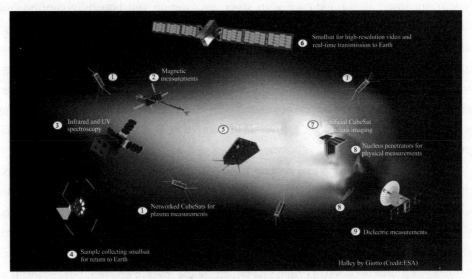

**Fig. 2. 3**    Example of concept meant to explore comet Halley using a fractionated constellation during its next visit in 2061. This figure illustrates the types of measurements needed to fully understand this unusual comet (it does not belong to the two classical reservoirs of comets, Jupiter Family and Oort Cloud). Image Credit: ESA

there has been a tremendous development in regards to the technology maturation level achieved by SmallSats (NASA, 2015). Smallsats offer a number of advantages, in particular advanced distributed spacecraft architectures that can be used to address the above challenges and enable wholesome science investigations over a short observation window. This includes: ① a loose coordination to synthesize a single, large, virtual instrument (Bandyopadhyay et al., 2016); ② innovative distributed, possibly heterogeneous measurement, and data analysis techniques; ③ autonomous operations; ④ communication relay strategies; ⑤ novel orbital organization approaches for constellations or more effective swarming and to enable observations from multiple vantage points.

We (collectively) do not know how to approach objects with velocities in excess of 50 km/s. The Halley comet missions, while bold, had a modest science return in comparison to the level of resources engaged, because the violence of the heavy dust environment destroyed some of the instruments. However, these missions were milestones that sparked the development of miniaturized instruments in Europe and Japan's line of science missions. Similarly, we expect that objects of major science significance like debris from extrasolar planets and pristine building blocks of our solar system can foster novel approaches to space exploration and hopefully coordination among space agencies. A major aspect of this type of concept targets technology challenges related to manufacturing and operating large numbers of assets, resilient approaches to handling risk, and defining an effective framework to engage prospective sponsors, possibly from the international community. Private companies with internal and government support are paving the way for large-scale manufacturing of capable space platforms at low recurring costs, and offer a business model that could be a model for future endeavors.

It is envisioned that up and coming telescopic facilities such as the Pan-STARRS2 Observatory combined with the Pan-STARRS1 telescope, and later the Large Synoptic Survey Telescope (LSST) when it comes online in the early 2020s, will enable the discovery of LPOs a decade and more before these objects reach perihelion. This timeline is a priori sufficient to implement and launch spacecraft that may encounter the LPO as it approaches its perihelion, which should involve crossing Earth's orbit in most cases. Clever mission design frameworks need to be thought out ahead of time to address the aforementioned challenges.

## 2.2.2　Discovering exoplanets

This idea follows in the footsteps of the ASTERIA[①] 6U CubeSat Mission (Arcsecond Space Telescope Enabling Research in Astrophysics) that was successfully launched and deployed from the International Space Station in the Summer of 2017. ASTERIA is primarily a technology demonstration and an opportunity for training early career scientists and engineers. The mission introduces capabilities that enable long-term pointing and photometry monitoring at specific stars believed to host exoplanets. The main scientific objective of the mission is to search for transits of planets in front of their stars, expressed in the form of variations in the brightness of the latter. The capability to point for hours pertains to a large number of other astrophysical applications, for example to measure star properties. If ASTERIA is fully demonstrated, then it would make sense for a follow-on mission to send a large number of similar CubeSats, each of which would target a different star. The CubeSats may differ in the nature of the measurements they are performing, for example by carrying different filters (Cahoy, 2015). Slightly larger spacecraft may

---

① https://www.nasa.gov/mission_pages/station/research/experiments/2513.html.

allow for more complex techniques such as infrared or ultraviolet spectroscopy. For example, the recently selected SPARC mission (Star-Planet Activity Research CubeSat) is planning to assess stellar radiation environment via photometry monitoring in the ultraviolet (Shkolnik et al., 2018). The key to this type of concept is to dedicate one CubeSat per star target of interest. Thus, it makes it relatively easy for international collaboration once the concept of operations is agreed upon, i. e., everyone can launch as they see fit, and join different phases. It may allow for citizen science as well.

## 2.2.3 Giant planet magnetosphere and atmosphere exploration

This idea builds on the prospect that large missions to giant planets could have enough mass margin to carry several CubeSats that may be deployed in the atmospheres or magnetospheres of these planets. The icy giants Uranus and Neptune have been identified by NASA[1] as targets of prime interest for the next decade. Understanding the intrinsic magnetic fields and magnetospheres of these planets are important objectives of a future mission. Similar to Earth, giant planet magnetosphere characterization is best approached via multisite measurements. Preliminary analyses identified that simultaneous magnetic field measurements covering a broad range of latitudes and longitudes and pursued for at least a full rotation period (of the order of 10 h) would yield groundbreaking results in comparison to the current approach of this type of measurement. The CubeSats may be released sequentially for extended temporal sampling. High-quality magnetometers are small enough to fit within 3U CubeSats (see for example the INSPIRE[2] mission) and the latter may also include a transponder for gravity field measurements. This type of geophysical measurement is best realized if the CubeSats perform their measurements in a synchronized manner via telecommunication networking (among CubeSats or between CubeSats and mothership). Networking provides additional advantage, for example CubeSats flying by different hemispheres could perform sounding of the planet atmosphere via radio-occultation.

A different application could target planetary atmospheres where the deployment of many CubeSats in multiple sites would inform on chemical (e. g., volatile,

---

① https://www.lpi.usra.edu/icegiants/mission_study/.
② https://www.jpl.nasa.gov/cubesat/missions/inspire.php.

isotopes) composition and its lateral variations. That type of investigation would not require networking among CubeSats.

While there is strong interest from the community for this type of investigation at icy giants and other planetary bodies, the pathway for adding CubeSat-class spacecraft to flagship-class missions is not yet defined. A compromise may be sought where CubeSats are developed following design rules driven by the more expensive mission, with the risk that they might become too expensive for multiple of them to be carried in the first place. Approach to risk and mission assurance might also make it more difficult for a mission from a space agency to carry CubeSats developed by foreign entities.

There are a number of technology roadblocks that need to be addressed before these three or any other major deep-space missions can be undertaken with small satellites. Telecommunications, power generation, propulsion, and mission operations pose specific challenges different and more severe from missions in Earth orbit, as already discussed in Section 1.2.3.

> **Finding 2.2** Small satellites provide opportunities to significantly enhance infrequent interplanetary missions with, e.g., landers or sacrificial satellites, and networks of small satellites that could enable missions to "once in a lifetime" objects.

## 2.3　Small satellite synthetic aperture telescopes

Many of the primary scientific goals of astrophysics require making observations of the faintest objects in the universe and forming images of stars and planets with sufficiently high spatial resolution to resolve their disks. These are tasks for large collecting area and/or large effective apertures.

NASA's James Webb telescope, for example, to be launched in 2021 and costing some $9 billion, is about the largest practicable telescope that can be origami-folded into the largest available launcher fairing. A different approach will be needed for the next generation of telescopes if, say, double the aperture is required (Sweeting, 2018). Small satellites with mirror segments could either be assembled (or auto-assemble) to larger structures in space (see Fig. 2.4, Saunders et al., 2017) or even operate together in free formation flight.

Specialized constellations of SmallSats will in the near future be able to make synthetic aperture telescopes with both large collecting area and/or large effective apertures. These new generation telescopes would be able to image planets, resolve stellar systems, and detect and image near-Earth asteroids at costs that are significantly less than has occurred in the past.

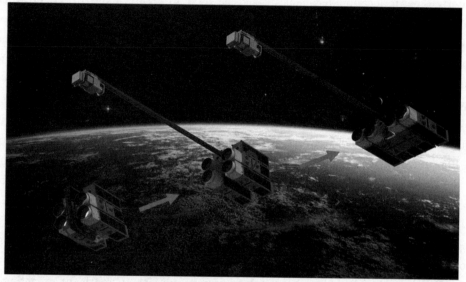

Fig. 2. 4    Concept of the Autonomous Assembly of a Reconfigurable Space Telescope (AAReST) mission consisting of an arrangement of 10 cm diameter circular mirrors attached to a cluster of CubeSats. Image Credit: AAReST Team, California Institute of Technology ( http://www. pellegrino. caltech. edu/ aarest1/)

For decades, radio astronomers have used synthetic apertures to achieve high spatial resolution and large collecting areas. As a consequence the methods for reconstructing images from distributed arrays of telescopes are well understood. Observatories on the ground by, for example, the VLTI and CHARA arrays have demonstrated visible synthetic aperture systems implemented by combining individual telescopes using beam directing mirror systems, evacuated tubes, and automatic phase delay controls. It is possible, though challenging, to apply the techniques demonstrated on the ground in space.

This is not a new idea. The Space Interferometry Mission (SIM) that started in 1998 was intensively studied, but ultimately dropped for technical and cost reasons. The Laser Interferometer Space Antenna (LISA) is underway as a joint NASA/ESA mission to detect gravitational waves. LISA requires pointing precision well beyond that of an optical synthetic aperture array. The LISA Pathfinder mission has flown and has exceeded its design requirements.

The lessons learned from the SIM and LISA Pathfinder missions together with new small atomic clocks, optical communications between telescopes, and precision interferometric location techniques could be used to create a distributed array of 200 kg one-meter telescopes. Sub-nanosecond clock synchronization can be achieved using pulsed optical links (Anderson et al., 2018). The collection area would depend on the number of telescopes in the array; an effective ten-meter telescope would require about one hundred spacecraft with one-meter telescopes, and a 30 m system would need about a thousand satellites. New manufacturing techniques being developed by industry for EO constellation missions could be applied here. Manufactured in quantity, a reasonable target cost per spacecraft could be ~ $ 500,000. So, the cost of a 10 m and 30 m distributed array could be ~ $ 50 M and ~ $ 500 M respectively. The cost of the design and development plan might be $ 100 M. The launch costs would be comparable to the cost of building the satellites, so it is not unreasonable to expect a 10 m and 30 m telescope to cost $ 200 M and $ 1100 M respectively. This is significantly less than the James Webb space telescope even if the cost estimates are low by a factor of three or more.

Developments in photonics technology provide another approach for executing a phased array optical telescope. New fabrication developments have allowed the construction of a 1.2 m flat panel phased array. The first mission could be a 1.2 m system on a small satellite. The next step might be to build a folded array of one-meter panels. Unfolding a stack of 9 planar arrays, a 3 × 3 panel telescope, would produce a collecting area of 13 m², equivalent to a 4 m telescope, with a spatial resolution of 0.028 arcseconds in the mid visible. The array panels could use small lenses on each of the photonic waveguides. Each lens could have a nano grating on its surfaces to generate a spectral shift with angle. Because the telescope is pointed electronically, images in different spectral bands could be obtained. Alternately, each wave guide could have a tunable photonics Fabry-Pérot interferometer for high spectral resolution. The elements of the array would be connected with a single-mode optical fiber to computer controlled photonics phased delays in the spacecraft's correlator.

Another interesting possibility is a spacecraft with a one meter photonics telescope on the spacecraft and another that is deployed on a long (e.g., 100 m) arm. This is challenging from both the mechanical and the thermal stability perspective even though the arm does not need to be rigid to optical wavelengths. Its role would primarily be to provide and keep the relative location of the telescopes. At some arm length the mechanical and thermal challenges will become more demanding than those associated with flying in free formation, but today it is anyone's guess at what scale the transition will occur.

A 100 m system would have a spatial resolution of about $5 \times 10^{-9}$ rad or a milliarcsecond in the mid visible. In a 1000 km near-Earth orbit, it could resolve a 5 mm feature on the Earth's surface, features on the Moon as small as two meters, and 19 m on an asteroid at 100 lunar distances, respectively. It could resolve features of $5 \times 10^5$ km on a star at ten light years. On nearby stars the system would be able to resolve starspots and apply the techniques of helioseismology that have been developed to determine the solar temperature, density, and rotation rate in the stellar interior.

It is very much hoped that the James Webb space telescope will live up to expectations and deliver groundbreaking observations in 2021 and onward. But it is equally clear that the next step after this can only be made with a distributed system that seems way out of reach technologically at this time. The situation is somewhat reminiscent of ESOs Very Large Telescope (VLT), made of four 8 m telescopes, that saw first light in 1998, but only recently did it become possible to combine their signals to make a single 16 m telescope. [1] In space, it will be small radio telescopes that will make interferometry possible first, and from there we can work our way through infrared into the optical band. In parallel, progress made in attitude control with the LISA Pathfinder and the forthcoming LISA mission will bring the necessary precision to combine optical signals so that it may eventually become possible to, e.g., image an Earth-like planet in another stellar system.

> **Finding 2.3** Monolithic large telescopes in space cannot grow further after JWST. A new approach such as distributed apertures on small telescopes is needed to make further progress.

## 2.4 Interstellar missions

Today, interstellar missions are impeded by the vast distances of space in combination with the limited lifespan of human beings. Thus, in order to enable future interstellar missions, the velocity of spacecraft must be increased, either by increasing the initial acceleration or the time over which acceleration is applied.

---

[1]  https://www.eso.org/public/news/eso1806/.

The concept of solar sails utilizes the solar radiation pressure for propulsion. This removes the need for carrying propellant tanks, thus reducing system mass and complexity. However, as the solar radiation pressure is very low, large sail areas are needed and very long acceleration times must be endured. The successful JAXA-built spacecraft IKAROS demonstrated the technology on a Venus bound mission launched in 2010 (Tsuda et al., 2011). The Planetary Society's LightSail-1 mission deployed a solar sail in orbit in 2015. [1] More recently InflateSail, launched in 2017 as one of the QB50 satellites, did the same with a 3U CubeSat (Viquerat et al., 2015).

A similar concept is the laser sail in which the propulsive photons are generated on Earth. One such mission concept is the Breakthrough Starshot initiative proposed in 2016 by Stephen Hawking, Mark Zuckerberg, and Yuri Milner. [2] This effort strives to lay the groundwork for a mission to Alpha Centauri within the next generation. The concept builds on two major ideas: shrink current-day spacecraft to a total mass of about 1 g and leave the propulsive system based on a laser on ground. They show that with such a system, spacecraft velocities of up to 20% of the speed of light are feasible (Lubin, 2016). With such high velocities, it will be possible to leave the solar system and conduct interplanetary missions within the average lifespan of space researchers.

The technical challenges of the Starshot initiative are formidable. The initiators admit that, "A number of hard engineering challenges remain to be solved before these missions can become a reality", and go on to list some 29 of them while claiming that, "no deal-breakers have been identified". For example, the feasibility of a sail with the required properties is far from clear as it requires managing multiple, conflicting priorities, and engineering a solution that partially satisfies all of them (Atwater et al., 2018). Moreover, the nanocraft will be subject to potentially damaging collisions with interstellar gas and dust (Hoang et al., 2017). The communication with Earth will suffer from a very poor link budget that will need to be addressed by either repurposing the sails for communications and/or distributing the satellites along the way as relay stations. Finally, the question of what kind of scientifically useful measurements could be obtained also remains. Even so, a mission to a star other than the Sun remains the ultimate vision for the future and is well worth exploring further.

---

[1]　http://www.planetary.org/explore/projects/lightsail-solar-sailing/.

[2]　http://breakthroughinitiatives.org/initiative/3.

## 2.4.1  Challenges and impact

The basic idea of leaving the propulsive system on Earth opens up a new class of missions and research projects not confined to interstellar missions. An Earth-based infrastructure for laser propulsion would also allow for faster interplanetary missions. It may specifically constitute the base for a fast response system for missions to unexpected targets.

A wide range of technologies will have to be developed before any such missions are launched. These include new energy systems capable of storing energy in the GWh range and delivering this energy to the ground-based laser system almost impulsively at bursts reaching 500 GW. But also powerful lasers, ultrathin sails with ultrareflective coatings, more energy-efficient communications systems, and new integration techniques for the actual spacecraft have to be developed. Many of these technologies may be used for terrestrial applications as well, improving society in general. For instance, the needed development of the laser power supply may lead to an increase of the efficiency of terrestrial power plants and distribution systems, or help in solving the energy storage problem that renewable energy sources have to tackle.

## 2.4.2  Pre-interstellar missions

Prior to interstellar missions, the mere testing of the spacecraft system and design within the solar system will allow for a new branch of scientific studies, such as:

 • A fast response system to explore the unexpected, e. g. eruptions on solar system bodies or the interstellar asteroid A/2017 U1 'Oumuamua (Gaidos et al., 2017).

 • 3D mapping of asteroid belt objects using a swarm.

 • Multi-point studies of the heliosheath and termination shock.

With velocities approaching a fraction of the speed of light, intrasolar system travel times are dramatically reduced. Further, once the development and construction of the necessary ground-based infrastructure is done, the launch cost is reduced to the maintenance cost of the Earth-based infrastructure and the energy required to accelerate individual spacecraft. This may open up deep space for a much more diverse scientific audience similar to what CubeSats have done.

The Earth-based laser propulsion system consists of a laser array and is thus fully scalable. This means that ramping up the accelerative force only requires that extra lasers are added to the laser array. This also allows for a trial-and-error approach to the mission scenarios. Fig. 2. 5 illustrates energy expenditure, energy cost[①](0. 12 € / kWh) and ultimate velocities of spacecraft with increasing mass being accelerated by the same array. It assumes a laser array of 10 MW with an array size of ~120 × 120 m² propelling spacecraft of masses from 1 g to 1 kg. Since heavier spacecraft accelerate more slowly, they remain within the vicinity of the laser for a longer time, which increases overall energy consumption and cost. The laser array size has been determined by extrapolating from values given by Lubin (2016).

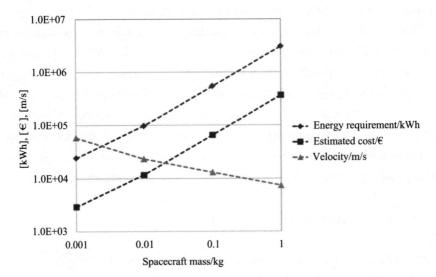

Fig. 2. 5   **Graph illustrating the relation between the spacecraft mass and laser array energy consumption, cost, and final velocity, in units given in the legend**

Small laser-propelled spacecraft are the only ones that could catch up with objects moving at tens of km/s and be cheap enough to be on standby in Earth orbit. Although this speed is far from the ultimate design speed of 0. 2 c (60000 km/s) for the starshot, it could represent a good intermediate step. If for example a 1 g spacecraft in Fig. 2. 5 is accelerated to ~1/1000th of the ultimate speed, roughly 60 km/s, it could have made a rendezvous with A/2017 U1 'Oumuamua at closest approach in about 8 days. That would require that 'Oumuamua had been spotted in time (which was not the case). If we assume that a hot pursuit was started at the time when 'Oumuamua

---

①   http://ec. europa. eu/eurostat/statistics-explained/index. php? title＝File: Electricity_prices_for_non-household_consumers,_first_half_2017_ (EUR_per_kWh). png.

was discovered it would have taken about a month to catch up. Fig. 2. 6 shows the trajectory of 'Oumuamua from a distance of 1 AU before closest approach to a distance of 12 AU after closest approach. Using the final velocity of the three spacecraft with masses of 0.001, 0.01 and 0.1 kg in Fig. 2.5, it is possible to calculate how far they travel daily. The distance from Earth is set to 0 AU until the day they are launched towards 'Oumuamua (which is assumed to happen on the 20th of October-the day after discovery). By calculating the distance between 'Oumuamua and the launch position, it is possible to estimate how long it will take for the spacecraft to catch up with 'Oumuamua, as illustrated in Fig. 2. 7. It is seen that the 10 MW laser array is not capable of accelerating the 0.01 kg and 0.1 kg spacecraft to a final speed that will allow them to catch up with 'Oumuamua. Either a more powerful laser array or a better sky survey system that provides earlier alerts will be needed. The energy cost at today's European electricity prices would amount to approximately 30000 € per 1 g spacecraft accelerated.

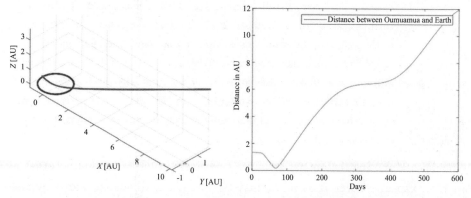

Fig. 2. 6　Left: the trajectory of 'Oumuamua as it passed Earth orbit. Right: the overall distance between Earth and 'Oumuamua. Position data used for the illustration is taken from JPL Horizons database and web interface (https://ssd.jpl.nasa.gov/horizons.cgi). 'Oumuamua was discovered after closest approach

After the acceleration phase, the "ChipSats" are coasting, which means that any orbit perturbations must be a result of external forces. By simply tracking the chipsats, the nature of such forces may be studied. Adding for instance a magnetometer to the spacecraft would allow the study of the space traversed while keeping the data return scalable. Imaging instruments will further enhance the capability of the spacecraft but also increase the demand for data bandwidth. In the proposal by Lubin (2016), the laser array is used both for acceleration and for

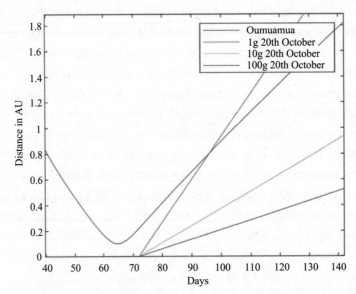

**Fig. 2.7** The blue line shows the distance between 'Oumuamua and the position of Earth at the 20th of October 2017. In this simulation the S/C are accelerated from that point in space and at that date. The three lines (red, yellow and purple) indicate three different spacecraft masses all accelerated with the same 10 MW laser array. It is seen that only the 1 g spacecraft will gain sufficient velocity to catch up with 'Oumumua (For interpretation of the references to color in this figure legend, the reader is referred to the web version of this article)

communications using the optical array as a receiver. Alternatively, a network using other deep space spacecraft as relays may be envisioned, similar to the relay system demonstrated by NASA using its Mars orbiters to communicate with rovers. The two MarCO spacecraft launched with InSight demonstrated a CubeSat version of such a relay system in November 2018.

## 2.4.3 Politics

The infrastructure for interstellar propulsion requires large investments. Laser arrays are envisioned to cover from $10^6$ to $10^8$ m$^2$ depending on the power output, which shall be between 1 and 100 GW. In terms of size the construction is comparable to the LHC at CERN (8.6 km in diameter) and much larger than ESO's ELT,[1] both of which are the results of international collaboration and funding. Though no cost estimates have been made, it is likely that a number of space entities need to partake in the construction in order to secure the needed funds. The initial Breakthrough Starshot proposal suggests a large array of lasers to be placed at one location, selected from a mission requirements point of view. However, it might be easier to obtain the required funding for the structure by dividing the array into smaller entities placed at locations chosen from both a national political point of view as well as mission requirements. Spreading the array over multiple nations will also, by necessity, strengthen the international space collaboration just as the international space station has done. However, distributed laser arrays may introduce laser phasing issues.

When operating, the laser array emits laser power in the GW range, thereby endangering any object that happens to be in the light beam, including satellites orbiting Earth. Depending on the mission type and spacecraft size, operations may span between ten minutes and several hours. With more than 40000 objects in orbit around Earth, close coordination will be paramount to ensure safe operations. Conversely to the hazardous potential, the propulsive power of the system may also serve purposes of international relevance such as collision avoidance of low Earth orbit spacecraft (Stupl et al., 2012) and asteroid deflections (Thiry and Vasile, 2014).

## 2.4.4 Technology

Developments in CMOS technology and space technology in the form of CubeSats have shown that shrinking physical size and mass is possible while maintaining most or all capabilities of a system. Though the Starshot mission is based on available technologies, many of these are either not yet adapted for deep space missions or are still at a low technology readiness level (TRL). To date, the lowest mass tech demonstrators are the Sprite satellites based on one single printed circuit board. They have been launched on two occasions, but have not yet performed separately, i. e. detached from the mother spacecraft.[2] Thus a suite of enabling technologies needs to be either transformed or developed further to facilitate the Starshot mission.

---

① https://www.eso.org/sci/facilities/eelt/.

② https://www. scientificamerican. com/article/reaching-for-the-stars-breakthrough-sends-smallest-ever-satellites-into-orbit/.

The envisioned laser wavelength is 1056 nm, i. e. in the IR regime. Though Earth's atmosphere is fairly transparent at this wavelength, high grounds such as mountain regions would still be preferable locations. In the fully developed system, Lubin proposes an ultimate laser array with an optical power on the order of $1\sim100$ GW. Typical wall plug efficiency (WPE) of lasers is around 20% (Botez et al., 2015), though 50% have been reported (Pietrzak et al., 2015). Thus for the largest single site system a nearby power plant capable of delivering up to 500 GW over a period of 10 min would be required. For comparison, a large nuclear power plant typically have a power output of 2 GW[①] and the Space Shuttle outputs 45 GW at take-off. Delivering power to such systems is clearly one of the major infrastructural challenges which will require substantial political support in the country or region in question. The enormous, almost impulsive energy requirements could be somewhat relaxed by dividing the array into minor subarrays. This will introduce a new challenge of phasing the laser beams, though.

## 2.4.5　Predicting when to launch —The technology race

The need for miniature interstellar spacecraft will either introduce or accelerate a spacecraft mass reduction trend. It is possible that such a trend has already been instigated by the advent of the CubeSat format. Here we attempt to estimate the rate of such a trend by combining the predictions given by a spacecraft version of Moore's law with Newton's second law, which gives the travel time $t_c = (dm_{s/c})/(t_a F)$ that it will take a spacecraft of mass $m_{s/c}$ to reach a star at distance $d$ if accelerated initially by a force $F$ during a time $t_a$.

To the best of our knowledge, there are no in-depth studies of a mass reduction rate for spacecraft. The mass reduction rate is the time it takes before a given performance may be obtained at half the mass. Two smaller surveys have been conducted, one looking at 115 missions divided into four mission types (Fléron, 2017) and another conducted on 47 communication satellite missions (Fléron, 2018). The first study indicated a mass reduction rate of 36 months for Earth observing missions, but showed no apparent trend for the three deep-space mission types. The second, studying communication satellites, indicated an 18-month mass reduction rate, which seems very high but due to the relatively small data set this may be inaccurate.

---

① https://en. wikipe dia. org/wi ki/List_of_large st_power_ stations＃Nuclear.

Another way of estimating the mass reduction rates is by looking at the simplest satellites that only emit a beacon. With only one function they are easy to compare. Sputnik, Cute-1 and the Sprite satellites were such beacon satellites (Fig. 2.8).

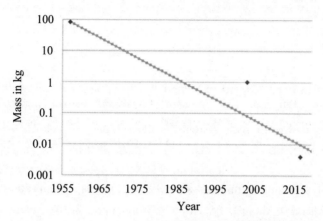

Fig. 2.8　The mass evolution of beacon satellites Sputnik (1957), Cute-1 (2003), and Sprite (2017). The grey trend line has a halving rate of 55 months

Assuming that CubeSats have instigated a Moore's law for spacecraft as indicated, then the mass reduction rate may be used to calculate the optimum launch time for a given mission. To illustrate the process let's assume a mass reduction rate of 4 years and an initial mass of 1000 kg in the year 2000. All missions are assumed to use the same Earth-based laser array for propulsion as defined by Lubin (2016). Thus the accelerating force, $F$, remains the same throughout the years. The 1000 kg mass of the first spacecraft mass was chosen arbitrarily inspired by the New Horizons spacecraft of 478 kg launched in 2006 and the Juno spacecraft of 3625 kg launched in 2011. It also corresponds roughly to the mass of the Voyager probes, which were launched in the 1970s and today are the first interstellar probes (but will take 75000 years to reach the distance of the $\alpha$ Cen system). As a result of the expression for the travel time above, the ChipSats and similar technology will go deeply interstellar first. Even if interstellar missions using larger crafts are launched in the near future, these will be overtaken en route by the ChipSats simply because these will attain higher coasting velocities, see Fig. 2.9.

This last remark should not be mistaken as a recommendation to just wait until ChipSats and the laser launch system are mature enough before attempting an interstellar mission. In fact the first interstellar mission, Voyager 1, although not declared as such, was launched already in 1977, as mentioned above. Currently, an

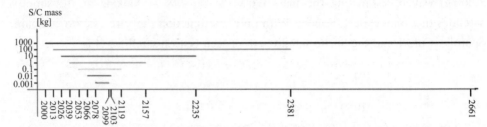

Fig. 2. 9    The launch and arrival times (at Proxima Centauri) as a function of the
spacecraft mass. The evolution of spacecraft mass is assumed to follow a
spacecraft version of Moore's law. Here the half time for spacecraft mass is set
to 4 years (Fléron, 2017). Data for spacecraft velocities have been taken from
Lubin (2016)

Interstellar Probe mission is under study, although not as a small satellite, that should
be able to reach 1000 AU in fifty years (McNutt et al. , 2018), about six times faster
than Voyager. The COSPAR Panel on Interstellar Research (PIR) draws upon recent
and ongoing studies of the requisite science and miniaturized instrumentation
technologies to lead to an international consensus on approach and implementation of
such missions. Even though the ultimate goal of reaching another star seems far out at
present, any attempt of getting there will help pave the way and lead to technological
advances and scientific discoveries.

> **Finding 2.4**   Engaging in exciting visionary goals such as Starshot, even if it
> turns out to be impossible in the end, will require us doing new things and
> developing new ideas and technologies that will have many applications by
> themselves.

# 3    Obstacles to further development and progress, and ways to overcome them

In this section, we address the obstacles between the current state of affairs
described in Section 1 and the visions for the future in Section 2, and try to identify
ways and means to overcome them. Here we concentrate on institutional obstacles as
opposed to scientific or technological ones: How can the scientific community, the
space agencies, industry, and the policy makers (governments and international

organizations) cooperate in a way to maximize the return of small satellites for space science?

Perhaps the biggest limitation to scientific progress in using SmallSats for science is imposed by high mission cost (especially launch cost, discussed in greater detail below), which limits the number of science missions and leads to a risk-averse posture of agencies necessitated by large and expensive missions. This in turn may stifle innovation by scientists who may become risk-averse in their proposals because of the review process and limited proposal opportunities. These issues could be mitigated by taking advantage of developments in the commercial sector and by increased international collaboration which distributes cost and risk. While there are additional difficulties with international collaboration, including timing issues in the decision making of missions involving more than a single agency, differences in their legal backgrounds, and others, a core challenge remains lack of coordination across these three sectors.

Government agencies have critical roles to play in both supporting utilization of small satellites, as well as promoting approaches that do not hinder innovation. Also, agencies are in a position to advocate for science-friendly policy decisions. They are the primary means by which, through advocacy, regulations that hinder greater scientific utilization of small satellites can be addressed. They also represent the primary mechanisms for leading and participating in multinational collaborations. Increased support for such collaborations could encourage the entry of new actors (government agencies acting as centres of technical know-how and providing technical service), connecting them with mission developers, demonstration missions, and institutional users.

Industry plays an important role in advancing areas important for science. For example, the increased availability of low-cost, high-reliability parts could bring down the cost of small scientific satellites. Lessons from industry in large-scale manufacturing and testing of small satellites could help make scientific constellation missions feasible. New models for low-cost access to space, such as the development of small launch vehicles and launch brokers may also drive down the launch costs of science missions.

Finally, there are cultural barriers preventing the full potential of scientific satellites from being realized. The culture within governments and the scientific community doesn't fully value small satellites. Their development and management approaches tend to emphasize low risk and high reliability, yet the culture in which SmallSats will thrive is one that allows for experimentation, risk-taking, and failure.

This chapter is divided into five sections. We first discuss the role of government agencies in supporting the development of small scientific satellites. This includes addressing issues related to funding and policy. We then discuss how both the science community and government can leverage developments in industry, and address the cultural changes required to fully realize the scientific potential of SmallSats. Lastly, we discuss collaborative models that can further the development of SmallSats for science and produce the robust workforce needed for future innovation.

# 3.1  Funding

Government agencies are the primary provider of funding for important science missions and for promoting development of the technologies that enable them. Scientifically-motivated agencies in the United States that play leading roles in utilization of small satellites are NASA and the US National Science Foundation (NSF), the latter restricted to CubeSats. It is noteworthy that the NSF took the bold lead for the support of scientific investigations utilizing the CubeSat standard as far back as 2008. By providing modest funding for its CubeSat initiative ($\sim$ \$ 900 k for each three-year mission), [1] NSF supported a fledgling community at a critical time. One can argue that this willingness to take a risk and develop a totally new program, helped pave the way for the scientific CubeSat revolution. Funding through NASA's Science Mission Directorate for scientific CubeSats has increased significantly, and proposal calls for Explorer missions and Missions of Opportunity now allow for CubeSat-based missions concepts.

Other U. S. agencies such as the National Oceanic and Atmospheric Administration (NOAA) and various organizations within the Department of Defense are using or exploring utilization of very small satellites to accomplish their missions. Notably, while not yet deploying government-owned nanosatellites, NOAA is in the second phase of awarding contracts to purchase climatological, atmospheric, and land imaging data from commercial providers who gather data from constellations of nanosatellites. [2]

---

[1]  https://sites. nationalacademies. org/cs/groups/ssbsite/documents/web-page/ssb _ 166650. pdf. Retrieved February 23, 2019.

[2]  https://www. nesdis. noaa. gov/content/noaa-continues-push-toward-innovative-partnerships-second-round-commercial-weather-data. Retrieved February 8, 2019.

In Europe, ESA is the main player and has been involved in small satellites since the 1990s. Funding opportunities for SmallSats have been somewhat sporadic, as described in Section 1. 1. 1. There are also efforts of individual nations (e. g., SwissCube) and of the European Union, specifically with the FP7 and Horizon 2020 program. There are already existing initiatives of national space agencies to support the formation of international partnerships such as Venls, an Earth observation and exploratory mission of the Israel Space Agency (ISA) and the French space agency (CNES), or the SHALOM mission, a joint initiative of ISA and ASI-the Italian Space Agency (Section 1.2.2).

On a global level, Appendix E of the IDA report on Global Trends (Lal et al., 2017) gives a comprehensive list of international small satellite activities and trends. It should be noted, however, that some of the countries so far focus more on industry than on science using SmallSats, e. g. the INDIA/ISRAEL@75 program, cf. Section 3.5.3. Moreover, as described in Section 1.1 and shown in Fig. 1.2, most SmallSats are launched by only a few countries. Thus, the use of SmallSats for science is largely undeveloped in most countries.

In addition to funding complete science missions, there are other areas where government investment can play a critical role in promoting the use of SmallSats for science. For example, government agencies can provide mechanisms for the development of cross-cutting technologies that enable more sophisticated space-borne capabilities in smaller packages. According to Lal et al. (2017),[1] since 2013, NASA Space Technology Mission Directorate (STMD) has invested about $80 million in SmallSat programs, primarily towards development of constellations, communications, mobility and propulsion. Notably, 60% of the funds went to industry. Other non-science US government agencies (e. g., Department of Defense) have also invested significantly in technology advancement that could benefit science missions. Lal et al. (2017) also identified specific areas in which government agencies should invest. These include "pre-competitive" R & D in areas such as mobility and propulsion, constellations and autonomy, thermal control, communications, deep-space systems and avionics, deployable systems, debris mitigation and control technologies, and others that science users consider important. They also include investment in risk reduction (i. e., by providing opportunities for on-orbit demonstration missions). And lastly, they include

---

[1] https://www.nasa.gov/sites/default/files/atoms/files/nac_march2017_ blal_ida_sstp_ tagged.pdf. Retrieved February 22, 2019.

investment in what are called "industrial commons" (i. e., shared knowledge and capabilities) in areas such as reliability testing and data curation. In creating industrial commons, governments should, to the extent possible, leverage existing organizations and learn from successful models.

Agencies are also in a position to enable frequent consolidated launch opportunities for small satellites, thus providing new opportunities for proof-of-concept and dedicated science and application development missions, with ambitious time to launch timeframes (e. g., 3 years from inception to launch). Government agencies can also promote the development of mission concepts to address observational gaps and ensure continuity of critical space measurements such as those required in Earth Observation.

The desire of agencies to impose technical standardization is a common theme. If setting standards, agencies should be mindful that depending on the specific circumstances, standardization can both benefit and hinder technological advances. Agencies are encouraged, if setting standards, to be loose, setting only the most necessary standards, and keeping those broad and flexible so as not to hinder innovation. By not imposing innovation-stifling programmatic constraints, agencies can take actions that promote free and open competition that favorably advances the capabilities of small space-borne systems.

Funding mechanisms other than the traditional national space agencies are emerging. Universities and university consortia, similar to those formed for building large ground-based telescopes, private foundations, private donors, and Kick Start projects are now developing space missions larger than CubeSats. [①] However, most funding for SmallSats remains in the United States, and some in Europe (principally the United Kingdom). This is likely because most other countries are focusing their space-oriented resources to developing operational communication and Earth observation systems rather than science. Given the low cost of SmallSats, it is feasible to jump-start space science programs without significant investment.

---

① e. g., ASU-Milo Project, MethaneSat-Environmental Defense Fund, and BeyondGo, a kick start.

**Finding 3.1**   There are specific areas that governments should support that the private sector is not likely to. These areas include technologies such as mobility and propulsion, constellations and autonomy, thermal control, deep-space systems and avionics, and debris mitigation and control technologies. One critical gap is support for "commons" and infrastructure technologies—activities such as database curation, facilities for reliability testing and launch support. The government should also more actively participate in activities that require coordination across the community, standards being one such area.

**Finding 3.2**   Set-aside funding mechanisms to support scientific SmallSats are needed, particularly outside of the US and Europe. Given that SmallSats have been shown to be useful for scientific advancement, there is a benefit in creating funding streams that specifically support small scientific satellites.

## 3.2   Role of policies that support the growth of small satellites

It is not just technology developments and government funding that would improve the alignment of small satellites with scientific use. There are several policy impediments that need to be addressed to ensure better use of small spacecraft for science. In this section, we discuss four that we believe are the most critical: access to spectrum; export controls; low-cost access to launch; and restrictions related to orbital debris.

### 3.2.1   Spectrum access

Electromagnetic spectrum for data transmission to Earth as well as accessing specific deep space bands is critical for any activity in space, and a scarce resource, at least for the time-being (until laser-based communications become the norm). As a result, access to spectrum is carefully coordinated and regulated at domestic and international levels, and it is illegal for any space object including small satellites to emit any type of radio signal without authorization. The framework for how the radio frequency spectrum is used is outlined in the Radio Regulations treaty ratified by the Member States of the International Telecommunications Union (ITU). Within that international framework, countries manage their national use of the spectrum. In the United States, for example, for small satellites owned and operated by NASA or other science agencies, the National Telecommunications and Information Administration (NTIA) typically grants authority to use a frequency [though does not issue a license, which is done by the Federal Communications Commission (FCC)]. When frequency usage is approved, the FCC and NTIA submit their frequency assignments to an FCC liaison, who submits them to the ITU, which maintains an international register. Scientific satellites have dedicated spectrum in all countries that have space programs. However, most scientific CubeSats, including NSF-supported CubeSats, have so far used frequencies in the amateur band with an amateur license. This is becoming more difficult and will likely not be possible in the future. Scientific satellites must now obtain an experimental FCC license and it is not currently clear which frequencies are accessible to grant-funded, university owned and operated CubeSats.

While spectrum-related issues are not qualitatively different for small satellites as compared with larger ones, the speed with which small satellites can be developed and launched is outpacing the ability of the current coordination process for spectrum allocation and management. In the United States, the FCC has been inconsistent and erratic in granting licenses, sometimes providing approval for a design and form factor, but sometimes not, even when the design is identical. [1] The challenge is exacerbated for international and joint projects where spectrum systems of multiple countries may need to be aligned.

There are other challenges as well. The expected large growth in small satellites[2] will place increasing pressure on the establishment for coordination in UHF, S, and X bands as well as other space allocated bands, since many commercial operators use spectrum that is being used or could be used by university or Federal government agencies. As more satellites are launched, the competition for spectrum would get even more intense, not just among satellites in LEO but also with satellites in GEO (for example, LEO satellites crossing the equator will have to change bands to avoid interfering with the GEO satellite, whose frequency rights take precedence). The

---

[1] See more at https://spectrum. ieee. org/tech-talk/aerospace/satellites/the-fccs-big-problem-with-small-satellites.

[2] By some accounts (e. g. , Aerospace Corporation reports) in the next decade, we may see up to 20000 satellites launched in LEO, most of them under 500 kg. However, accounts vary. According to Northern Sky Research, fewer than 4000 satellites are likely to be launched in this timeframe, and according to Euroconsult, more than 6500.

shared use of spectrum involves conducting interference analyses and extensive coordination, most of which must be completed as part of the regulatory licensing process. This is an additional element of risk for small satellite developers. Also, as RFI becomes more of a problem, an enforcement of current national and international mechanisms to regulate radio frequency to prevent interference might further step up, challenging the science community to continually stay apprised of changes to the system. On the other hand, it is also critical for ground-based radio astronomy that protected bands remain protected. Thus such regulations are good for science as well.

There are other issues that small satellite operators must face just the same as traditional satellite operators, some of which relate to competition between spectrum use for space-based versus terrestrial application. Several bands under consideration for terrestrial use (e. g. , spectrum above 6 GHz as part of 5G growth) are adjacent to critical bands such as those used for remote sensing from space. As a result, degradation in ability to use these bands is a growing concern (Mistichelli, 2016).

There are some issues specific to the SmallSat community. The procedure for receiving permission for spectrum use is long, complicated, and in many countries, spread across multiple agencies. Most researchers working on science small satellites are typically unfamiliar with these roles and regulations, and sometimes discover too late in the development process, and risk getting denied a license. Small satellite developers typically favor lower frequencies, where equipment is less expensive and more readily available, but lower frequencies are the most congested parts of the radio spectrum. The increasing use of small satellites may increase the need for higher bandwidth which has its own set of costs and challenges. Regulatory authorities also prefer to know details of satellite orbits when filings are made, but these parameters may be uncertain for some researchers until late in the process, for example when they find out who would launch their satellites as a rideshare.

## 3.2.2 Export control

Most countries have laws and regulations in place to protect the acquisition— especially by entities or countries they consider adversaries—of technologies or products that they believe safeguard their national security and foreign policy objectives. Space technologies and products comprise critical subsets of these technologies and products as they are almost always dual use, in that most space-oriented technologies and products, even when designed and used for science, can in principle be used for military purposes.

While export control regulations typically do not apply to general scientific, mathematical or engineering principles in the public domain (typically basic and applied research), they are often hard to interpret by university-based and other scientific researchers. In some countries, concepts such as "deemed exports"—which refers to items or information provided to a foreign individual—are often difficult to understand and adhere to, and the responsibility for complying with these laws often resides with faculty members and students not trained in such matters.

There is ongoing debate between government and academia regulated by export control regimes regarding the extent to which these restrictions harm legitimate scientific activity. Institutions of higher education in the United States argue that overly hawkish export control regulations could inhibit the best international students from studying in the US, and prevent cooperation on international projects. Over the years, export laws and regulations have become more complicated, and more aggressively enforced by government agencies. In the United States where this information is publicly available, university personnel have been prosecuted for breaches. Harmonizing international collaborations while ensuring export compliance of their research is becoming a precarious balancing act for scientists. In some cases, this can discourage scientists from participating in international collaborations.

It is also unclear if overly intrusive export control regulations are necessarily in a country's best interest. According to the inventor of the CubeSat, Bob Twiggs, a former professor at Stanford University's Space Systems Development Lab, "ITAR (or International Traffic in Arms Regulation, the US system of export control) is driving research out of the United States, isolating the United States and causing markets to be developed outside of the United States. Foreign students who are at the cutting edge of GNSS, electronics, control systems and rocket systems cannot do research in the United States."[1] The Executive Secretary of the US National Space Council has said that "burden(some) and outdated parameters can have the unintended effect of compromising national security by incentivizing space industries to move overseas, and for manufacturers to change their supply chain."[2] The same rationale applies to the scientific small satellite enterprises as well.

---

① From: https://www. satellitetoday. com/telecom/2008/08/01/itar-bal-ancing-the-global-playing-field/.

② From: https://www. hudson. org/research/14341-full-transcript-space-2-0-u-s-competi-tiveness-and-policy-in-the-new-space-era.

Small satellites provide unique opportunities for collaborations across nations, far more so than more traditional space activities. While small satellite projects do avoid many of the stringencies of export control regulations because of their use of COTS and other mainstream components and minimal use of sensitive technologies, there are no formal exclusions for small satellites for science use.

Clearly, in principle, export controls serve a useful function in improving national security by protecting against the transfer of critical technologies that should not be transferred. However if they include provisions that are too strong in preventing knowledge transfer, it can be detrimental to the country's long-term national security, in that it hampers the free exchange of knowledge that is essential to the success of space research in an increasingly globalized scientific community. Any export control regime that governs small satellites needs to find the balance between enabling progress of science including through robust international collaborations without impairing national military and economic development interests. This balance will come from a regular and robust dialog between space researchers using small satellites and policymakers regarding making periodic changes to the lists that include technologies of interest to the small satellite community (Broniatowski et al., 2005). For the SmallSat scientific community in particular, providing better clarity on the rules and regulations, including clearer interpretations, will go a long way in ensure adherence not only to the letter of the law, but also its spirit.

## 3.2.3　Access to space

In the past, small satellites have typically been launched through one of three principal ways: obtaining a rideshare on a rocket with a primary payload, such as a satellite or cargo for the ISS; ridesharing with a group of other small satellites on a "cluster launch" as was the case of the launch of 104 satellites on the 2017 PSLV launch; and buying a dedicated small launch vehicle, such as Orbital ATK's Pegasus rocket. Most launches of small satellites to-date have been as secondary payloads.

In the United States, NASA supports science SmallSats through the Educational Launch of Nanosatellites (ELaNa) Program under the CubeSat Launch Initiative (CSLI), and also subsidizes launches on commercial (such as cargo resupply launches to the ISS) and other (the EM-1 flight of SLS is expected to have 13 science and technology CubeSats) launchers. Scientists have other support options as well: outside the government, United Launch Alliance (ULA) provides competitive free rides for university-based CubeSats. ① To enable making connections with launch providers, companies like Spaceflight Industries and TriSept Corporation act as brokers for launch coordination and integration.

Today, globally there over a 100 launch companies dedicated to SmallSat launch. ② While a large portion of these launchers may not come to pass (most are in the development phase), the sector shows a dynamicism not seen in the SmallSat or launch communities in the past. For science users, a plethora of viable (and low cost) options for access to various parts of space is welcome news.

Despite the opportunities, there is a pent-up demand for affordable launch for scientific small satellites. The NASA CSLI has 38 CubeSats manifested and 66 launched of 162 selections as of December 1, 2018. ③ The Venture Class Launch Services (VCLS) program under CSLI is reducing the backlog via manifest of CubeSats on dedicated launch vehicles such as those offered by FireFly Space Systems, Rocket Labs, and Virgin Galactic. The first launch under the VCLS program was conducted by Rocket Labs on December 16, 2018, launching 13 CubeSats into Low Earth Orbit (10 under the VCLS program).

New launch options to serve the SmallSat community are emerging (e. g., Cappelletti et al., 2018); there are more than 50 companies developing small rockets to launch small satellites (Sweeting, 2018). However, rocket technology development is a notoriously high-risk enterprise, and many of these efforts are likely to fail. Moreover, small rockets tend to have a higher specific cost; while large rockets are expensive (Fig. 3.1), they offer the greatest economy as a rideshare option (Fig. 3.2). For example, the $62 million dollar Falcon 9 launches 22800 kg or $2720/kg which nearly a factor of 50 less that the cost to launch 1 kg on a Pegasus XL. On the other hand, rideshares offer the least amount of flexibility with respect to choosing an orbit or inclination of operation, or even the ability to have propulsive capabilities, factors that are important determinants of conducting good space science.

Nevertheless, launch remains a chokepoint for smaller satellites. If small satellites grow in number and utility as expected, low-cost launch availability will need to increase. While the announced large rockets such as New Glenn from Blue Origin, Falcon Heavy from SpaceX (which has already flown twice), and Vulcan from ULA

---

① http://www.ulalaunch.com/ula-reveals-transformational-cubesat-launch.aspx.

② https://www.spaceintelreport.com/count-em-101-new-commercial-smallsat-dedicated-launch-vehicles-in-development/.

③ https://www.nasa.gov/content/cubesat-launch-initiative-selections.

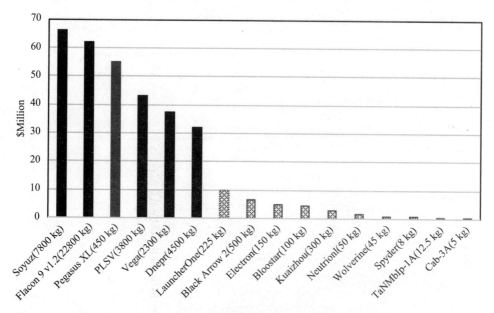

**Fig. 3.1** Price of launcher versus maximum launcher capacity, ordered by cost. Black bars represent large launchers and red bars represent small launchers; shaded bars represent launchers in development. Figure reproduced from Lal et al., (2017), with permission from the Institute for Defense Analyses (For interpretation of the references to color in this figure legend, the reader is referred to the web version of this article)

could support the SmallSat science community, they first need to address the issue of SmallSat integration.

A driver of low-cost access to space is the development and success of commercial space, as this would lead to a large increase of the launch flight rate, which could result in a low per-flight cost of launch. Depending on how the industry develops, it may also reduce opportunities for scientific SmallSats. Other factors (e. g., reusability) could also affect the cost of access to space. Until such a time, governments need to continue to subsidize launch of spacecraft for science applications.

## 3.2.4　Orbital debris considerations

Space is becoming increasingly more crowded, and the growth in the number of satellites in LEO is expected to be dominated by small commercial (not science)

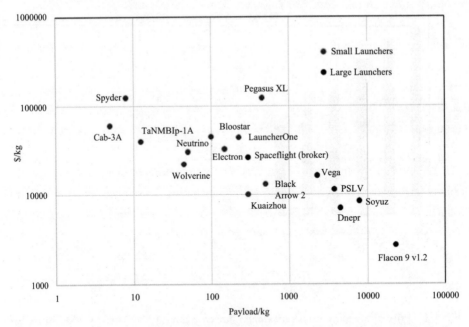

**Fig. 3.2　Price per kilogram for small and large launchers. Figure reproduced from Lal et al. , (2017), with permission from the Institute for Defense Analyses**

satellites. Varying accounts ascribe between 3600 – 6200 – 25,000 small satellites (satellites weighing less than 500 kg) to be launched between 2017 and 2026.[①] The concern is not just the increasing number of satellites, but also the fact that (depending on their altitude) they will stay in space as debris for longer than their useful life.

As the number of these spacecraft increases, the probability of collisions increases as well, especially if satellites are not able to be tracked well or are not maneuverable. Due to high speed in LEO (~10 km/s), even submillimeter debris poses a realistic threat to human spaceflight and robotic missions. Small satellites, and especially CubeSats, in near-Earth space are increasingly being seen as the major orbital debris challenge of the coming decades (Bastida Virgili and Krag, 2015; Matney et al. , 2017). Even though the fraction of scientific small satellites is small, the scientific

---

① If all of the more than 160 constellations—most of them leveraging small satellites-came to fruition (an unlikely scenario), there would be more than 25000 satellites in LEO. 90 percent of these satellites would focus on communications, and the remaining on earth observation and remote sensing.

community must contend with being viewed as part of this challenge.

Concerns related to collisions focus largely on the proposed "mega" communications constellations. A recent report from Aerospace Corporation evaluated the effect of adding two large constellations—that of SpaceX and OneWeb —to the current constellations in LEO (Iridium, Orbcomm and Globalstar) and found that within its first 20 years in orbit, the first constellation is expected to cause one collision annually; this number could grow to approximately eight per year at its peak collision rate, which occurs about 190 years after launch.

As the number of small spacecraft (especially in low Earth orbit) increases, there will likely be growing restrictions on operators, including for science, even though it is not spacecraft for science that are the root cause of the coming debris challenge. Restrictions are likely to be directed toward CubeSats rather than SmallSats in general because, up until recently, CubeSats typically did not have on-board propulsion, ability to be tracked if they were not actively emitting signal, or maneuverability. These restrictions are likely to address three areas in particular: ① ensure all small satellites can be tracked, either actively or passively; ② mitigate radio frequency interference (RFI) as discussed in the spectrum section above; ③ abide by stricter guidelines to deorbit after they stop functioning. On the last point, it is worth noting that in recent years, many experts have come to believe that the international guidelines that recommend CubeSats deorbit within a 25-year period after their operational period ends are no longer sufficient, and may need to be updated.

The science community has an opportunity to avoid potential future problems by continuing to proactively seek technological solutions, such as low-cost means for CubeSats to be maneuverable and trackable, avoid RFI, and deorbited in a timely way. More R & D may be needed to assess which are cost-effective.

## 3.2.5　Summary and findings

There are four key policy challenges that need to be addressed to enable effective use of small satellites for science applications. First, spectrum is a scarce resource, and the SmallSat science community needs to be much better educated about not only the process of obtaining spectrum allocation for their spacecraft (which can be time-consuming) but also emerging and fast-moving changes in the area. Second, export control laws of many countries inhibit scientific collaboration by putting an undue burden on scientists to ensure compliance with a complex system. Again, education of the scientific community is key here. Third, the cost of launch is a critical inhibitor of SmallSat-based science. Government agencies have typically subsidized launch, a practice that needs to continue in the near-future. It should be noted that cheap space access requires the large launchers. Small launchers dedicated for small spacecrafts have a higher per kg cost than the piggyback launches and hence serves a market segment that requires extended control over the orbit parameters. Last not but least, as traffic in space (especially low Earth orbit) increases, there will likely be growing restrictions on satellite operators, including for science. Restrictions are likely to be related to better tracking in space, frequency interference, and stricter guidelines related to deorbiting and debris mitigation. The science community needs to proactively address these challenges.

> **Finding 3. 3**　Spectrum access (for data transmission to Earth as well as accessing frequencies in bands for research) is critical for any activity in space, and a scarce resource.

> **Finding 3. 4**　The undue burden of complying with laws and regulations related to international exchange and collaboration are a deterrent to scientific collaboration.

> **Finding 3.5**　Low-cost launch, through easy access to rideshare options, has been a key enabler of SmallSat-driven science.

> **Finding 3. 6**　As traffic in space (especially in low-Earth orbit) increases, growing restrictions on small satellite operators, including for science, is likely. Regulations are likely to be related to tracking in space, maneuverability, and orbital debris mitigation.

## 3.3　Leveraging developments in industry

Funding for science still comes primarily from government agencies, but science can potentially reduce costs or increase capabilities by taking advantage of commercial efforts, particularly the emerging industries that focus on SmallSats. If the cost of a SMEX-class mission in near-Earth orbit could be reduced to the $25 million level, a factor of ten decrease, it would open up new possibilities for science and significantly increase the number of flight opportunities. Because it offers more frequent launch opportunities, the growing SmallSat industry can also help attract and retain talented scientists and engineers, helping to build the science and aerospace engineering workforce.

The development of the SmallSat sector was led by the private sector (including universities), and most SmallSats launched are by private or commercial organizations. As an illustration, in a database of over 650 SmallSat organizations, developed by Lal et al. ,(2017) over 50% of organizations globally, and over 75% in the United States are in the private sector. While in recent years, academic use of SmallSats has grown (Fig. 3. 3), commercial operators continue to dominate the sector. In the last six years alone, over 475 commercial SmallSats were launched (Halt et al. , 2019). Most commercial SmallSats are for remote sensing (see Fig. 1.3, Halt et al. , 2019), though it is expected that with the advent of commercial mega-constellations, more satellites will focus on broadband services from space. ①

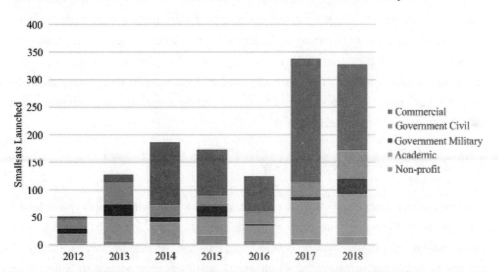

Fig. 3. 3    Number of SmallSats by sector. Image Credit: Bryce Space and Technology
(https://brycetech. com/reports. html)

There are several ways in which the science community can leverage the developments in the commercial sector. Increased access to space was already

---

①  e. g. , see projected small satellite launched by application at https://www. nsr. com/ smallsat-growth-on-shaky-foundations/.

discussed in Section 3.2.3 above. Here we describe ways in which science can partner with industry.

## 3.3.1 Commercial off-the-shelf (COTS) parts

In industry, commercial Earth observation and communications are the lucrative "killer-apps" of SmallSats, and comprise most commercial SmallSat activity occurring today. In these areas, the commercial sector is seeing not only a growing number of operators (companies such as Spire, Planet) but also a growing number of component manufacturers/suppliers (companies such as Gomspace, ISIS, and Blue Canyon, among others). This sector is focusing on mass manufacturing with the goal of decreased cost. While lower cost is important, even more important to the science community is the availability off-the-shelf flight-qualified parts. This trend is accelerating with the onset of large constellations (there are at least 16 companies focusing on use of constellations for earth observation or space-based Internet) that require at least the satellite bus (if not also the payloads) to be commoditized. Many of these companies are borrowing methods and technologies from non-space industries, for example, adapting parallax algorithms, similar to ones developed for automobile collision avoidance systems, to conduct SmallSat proximity operations. To further reduce cost, a number of manufacturers and operators are experimenting with COTS parts as inputs for their systems.

## 3.3.2 Commercial data buy

Private investment may also exceed (at least the unclassified publicly-available) government investment by an order of magnitude or more. As such, the scientific community should closely watch developments in the private sector, not only to procure products but also services. Commercial SmallSats may be able to collect data relevant to science. This is particularly true in Earth observation and space weather. NASA and NOAA have started small pilot programs to see if industry can produce data products to their standards. The stated hope is that this will lead to a state where the government can buy data rather than building and operating expensive satellite systems.

A developing area is weather forecasting. The value of weather prediction models that use radio occultation of Global Navigation Satellite Systems (GNSS) signals such as GPS has now been demonstrated. GPS occultation programs have been funded by Taiwan's National Space organization, NOAA, NASA and private entrepreneurs. Government funders wish to buy data to reduce cost of their forecasting operations and improve the accuracy of their forecasts (enabled by constellations because they provide a dense dataset and high revisit times). However, commercial developers may not be satisfied with just providing a data service to the government. Cooperation between government and private enterprise may allow data that are purchased and distributed openly by governments to be utilized by private industry to generate products tailored to specific customer needs. In particular, the value of near realtime customized weather data is increasing because of large computers and novel software systems. The idea of governments purchasing commercial data will thrive or not depending on whether the SmallSat developers find it more profitable to make products for commercial customers or products desired by the government agencies.

Current free and open data policies that exist for government-produced data sets are extremely valuable to the science community, leading to novel use of data for research. The open data policies also increase the data usage internationally. It should be emphasized that there is a risk of losing such open data policies if government-industry data-buy partnerships are pursued. Although commercial data opens potentially new opportunities, both the science community and government agencies must work to ensure that contracts are written in a way to preserve open data policies.

## 3.3.3　Hosted payloads

The Global Observations of the Limb and Disk (GOLD) mission is the first NASA science instrument to fly as a hosted payload on a commercial satellite. [1] GOLD was launched on 25 Jan 2018 aboard the SES-14 satellite that reached geostationary orbit in June. The measurements (Fig. 3. 4) will improve our understanding of the uppermost reaches of Earth's atmosphere, critical for understanding space weather. Because of its location at GEO, enabled in large part because it is a hosted payload, GOLD scans the entire Western Hemisphere every 30 min, enabling us to monitor day-to-day changes in the upper atmosphere for the first time. The increasing number of commercial constellations, especially in LEO, has the potential to provide new

---

[1]　http://gold. cs. ucf. edu/gold-will-revolutionize-our-understanding-of-space-weather/. Retrieved February 28, 2019.

opportunities for hosted scientific payloads.

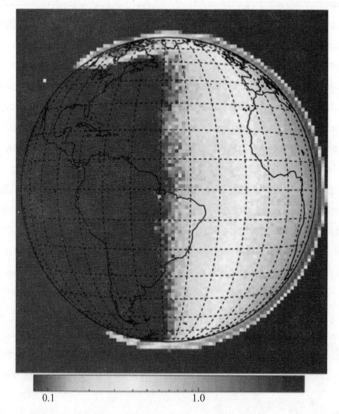

0.1                    1.0

**Fig. 3.4  First image of ultraviolet atomic oxygen emission at 135.6 nm from Earth's upper atmosphere captured by NASA's GOLD mission. Image credit: NASA/LASP GOLD Science team**

## 3.3.4  Industry-university collaboration

Finally, bringing industry to the table is an important step in creating an ecosystem of the SmallSat forces. When companies and universities collaborate to push the boundaries of knowledge, they become a powerful engine of innovation. The advantages to academia are self-evident: allowing students and researchers to work on groundbreaking research, greater potential for external funding, inputs for teaching and learning at the forefront of their disciplines, and the impact of providing solutions for pressing global challenges. Industry also benefits through workforce development and training and recruiting of potential future employees.

A recent example of such a partnership is NANOBED, a collaborative project led by the University of Strathclyde, in collaboration with Clyde Space & Bright Ascension, to develop a tool for research, innovation, and technology development. NANOBED enables rapid end-to end CubeSat mission design and technology development up to TRL6, including:

- Bespoke mission & system design software.
- Integrate hardware and test.
- Communicate with hardware over representative radio link.
- Simulate operations through day-in-the-life scenarios.
- Verify hardware functionality inoperational-like scenarios.

The mission and system design software allows the engineer to define a mission including orbit, ground station and spacecraft design, operational modes and switching conditions, and subsystem definitions from library or user-defined (XML) functions. The engineer can then simulate mission segments through orbital & attitude dynamics and control, ground track visualisation, power profile for each solar array and associated battery levels, ground station visibility, and data collection and downlink.

The software integrates with a hardware platform enabling an interface with flight hardware, including data acquisition units to check voltages, etc. The outputs of the mission & system design software can be used to drive power profiles to reflect solar arrays, and to invoke actions on hardware, including triggering ground station passes where GNU radio modules interface with radios for uplink and downlink. An image of the core hardware setup is shown in Fig. 3. 5, with space slots available for additional hardware such as payload units.

NANOBED has been deployed into a number of university and research institutes around the world, including in Mexico, the USA, and the UK, with others due for deployment into other countries, including South Africa, creating a global network of collaborators who can work together and share lessons learnt, both informally and formally through provided training courses. NANOBED provides a natural collaborative platform both within academia, but also between academia and industry.

The most fruitful form of cooperation is one that allows the participants to do new things that are hard or impossible to do themselves, can be built around a common research vision, and can continue for a decade or more, creating deep professional ties, trust and shared benefits that bridge the cultural gap between academia and industry. Long-term alliances build the human capital needed to make academia-industry cooperation work. Over time, a well-managed partnership produces an increasing number of professors and graduate students who can think and act across the

Fig. 3. 5　Core NANOBED hardware setup with flight hardware in laboratory at
University of Strathclyde. Image credit: Malcolm Macdonald

cultural gap, connect with the main research areas of the company and work in harmony to set joint strategic objectives.

> **Finding 3.7**　The availability of components off the shelf adds resilience and offers a fundamentally different way of building and operating scientific SmallSats. An approach with mass production techniques allows fast, innovative, and cheaper new space systems to be created.

> **Finding 3.8**　An increased number of commercial small satellite constellations may provide new opportunities for science through commercial data buy, hosted payloads, and ride shares. However, this comes with a risk that current open data policies could be in jeopardy.

> **Finding 3.9**　The SmallSat industry provides useful training grounds for students interested in aerospace science careers. Industry-academia partnerships, in particular, can help ensure a strong aerospace and space science workforce for the future.

# 3.4  Supporting innovation

The recent developments in the small satellite sector offer enormous opportunity for science. Realizing this potential to its fullest extent will require scientists and funding agencies to recognize that small satellites aren't just a miniature version of larger satellites. SmallSats can provide new kinds of measurements, be developed with a "fly-learn-refly" model, and can be lower cost. Indeed, it is not the satellite's mass that is its defining feature. The defining feature of a SmallSat is the unique culture that it engenders. This culture has more in common with a technology start-up that encourages risk-taking and rapid innovation, even at the expense of mission assurance, than with a traditional organization that emphasizes exquisite capability, long lifetimes, and high-reliability systems. New approaches that are specific to this organization are being embraced by industry, but there are cultural differences between these smaller and more nimble organizations and large space agencies[1] who, for the most part, still employ traditional approaches. For SmallSat-driven science to be at the forefront, a new paradigm will be required.

In "The Three Box Solution", Govindarajan (2016) emphasizes that different methodologies are required for addressing the three competing challenges faced by any organization: maintaining excellence in the present, identifying and letting go of outdated practices, and generating breakthrough ideas that can lead to future products or directions. In short, organizations that successfully manage all of these simultaneously do so by devoting entirely separate resources to each area. Innovation requires different skills, metrics, methods, and different management strategies.

Government agencies differ from start-ups, but the same principles apply. A report of the US National Research Council (2010) states, "... experience within DOD[2] has shown that actively managing basic research types of activities often requires different processes, metrics, and management techniques from those associated with managing advanced technology development and system prototyping activities." The same report points out that much of NASA's activities focus on risk mitigation which is necessary for large, expensive missions. On the other hand,

---

[1]  And also traditional space companies such as Lockheed which operate more like a space agency.

[2]  Department of Defense.

mission-enabling activities and innovation require a different strategy.

Given the unique culture in which SmallSats are likely to thrive, it might be useful to emulate organizational models where high-risk research—where failure is valued—is done. Several organizations, in the United States and other countries, have attempted to create a structure to conduct R & D that might be higher risk (with the likelihood of higher scientific payoff). These organizations could be models on how to nurture small satellites—by definition a higher risk proposition compared to traditional platforms—in larger organizations.

The Advanced Research Projects Agency (ARPA) organizations (e. g. , DARPA, IARPA, ARPA-E) in the United States have managed to create organizations where leadership prioritizes programs and projects that are high risk and not necessarily well-defined, seeking to maintain the integrity of the organization's high-risk culture (Peñ a et al. , 2017). [1] Similarly in the United Kingdom, the Engineering and Physical Sciences Research Council sponsors approaches within the larger organization (called IDEAS Factory) which aim to "stimulate highly innovative and more risk-accepting research activities that would be difficult to conceive under normal circumstances". [2]

One particular organizational construct to nurture the use of SmallSats in large organizations that will typically be resistant to the SmallSat culture is that of an island + bridge model that ARPAs use (Bennis and Biederman, 1997; Sen, 2014). In this model, the island is the refuge for experimentation and failure, and the bridge is the conduit for the transfer of knowledge and technology (to the user). New technological capabilities make their way out, requirements and other sorts of influence must make their way in. Research is neither entirely shut off from real-world interests, as with a traditional laboratory setting, nor is it beholden to the interests of operational incumbents. The island + bridge model applies a "connected science" approach to research, combining and integrating the forces of technology push and pull, balancing the need for isolation and connectivity, and providing just the right types of structure for processes that are necessarily chaotic (Sen, 2015).

Using this model, a SmallSat-specific sub-organization, created within the larger space organization, can make sure that SmallSats don't get short shrift, while ensuring connection with the larger organization (to make sure new technological capabilities make their way out and requirements make their way in). This model is neither new

---

[1]　https://www. ida. org/idamedia/Corporate/Files/Publications/STPI-Pubs/2017/D-8481. ashx.

[2]　https://epsrc. ukri. org/funding/applicationprocess/routes/network/ideas/experience/.

nor limited to the government. In industry, Lockheed Martin's Skunk Works and IBM's PC Project have shown the success of the island + bridge model.

> **Finding 3.10**　The culture in which SmallSats will thrive is one that allows for experimentation, risk-taking, and failure. Traditional space organizations tend to emphasize low risk and high reliability space systems, and it will take a difficult cultural change for such organizations to nurture SmallSats. There exist successful models in industry and even in government that can be emulated to ensure such a change.

# 3.5　Collaboration

The larger and more complex the project, the more critical it is to generate collaborative R & D to accomplish the goals of the project, for example in constellations with complex architectures or multiple means of measurement such as the EU QB50 or the TIM international projects described below. Although current consortia such as the QB50 are unlikely to survive the end of program funding, universities participating in a joint research program are much more likely to work together in the future. Increased engagement in multi-national scientific collaborations would also support the entry of new actors, connecting them with mission developers, demonstration missions, and institutional users. Such collaboration increases the chances of making a significant contribution to science while sharing resources and risks. It could also help to ensure that research tasks are dealt with by researchers with the most appropriate experience or with complementary interests and needs.

## 3.5.1　Models of collaboration

### 3.5.1.1　The TIM case study

Telematics International Mission (TIM)—is an example for which partners contribute individual satellites to a formation in order to benefit from the larger database generated. The low cost of very small satellites enabled, in this case, seven partner states of the Regional Leaders' Summit (RLS, e. g., the partner regions Bavaria, Georgia, Upper Austria, Quebec, São Paulo, Shandong and Capetown) to realize together a spacecraft formation for innovative joint Earth observation. ① TIM addresses a cooperating pico-satellite formation to generate 3D images for Earth observation by photogrammetric methods, taking advantage of the different viewing directions (Fig. 3.6). The obtained data will be fused for monitoring of environmental pollution, harvesting status, critical infrastructures, and natural disasters (like forest fires, volcanic activity, earthquakes).

Fig. 3.6    The three TIM satellites focus on the same target area for 3D-imaging by photogrammetric methods. Image Credit: Zentrum für Telematik (https://www.telematik-zentrum.de/)

The mission is currently in implementation stage with a planned launch in 2019. The scientific challenge relates to spacecraft engineering as well as to science data processing:

• Developing, in international partnership, modular, robust small satellites for a formation of networked, cooperating, "smart" small satellites, operating autonomously

① https://www.rls-sciences.org/small-satellites.html.

with minimum ground station interaction.

• Photogrammetric data processing for generation of 3D images, taking advantage of the large baseline distance between the instruments on different satellites and obtaining improved resolution by sensor data fusion methods.

Essential subsystems needed for a formation are: the attitude and orbit determination and control system, the communications system capable of inter-satellite communication and satellite-to-ground communication, as well as electrical propulsion for orbit control and maintaining formation. The core components are the 3 Bavarian satellites named TOM (Telematics earth Observation Mission). Each further contribution increases the formation capabilities and additional instruments provide complementary data. Precursor missions of the partners in this international team laid the groundwork of expertise in the relevant areas to enable this challenging pico-satellite formation flying application. Thus, through international cooperation, a challenging and innovative Earth Observation Mission is realized.

### 3.5.1.2 The QB50 case study

The QB50 mission demonstrated the potential of international university collaboration supported by the "new space" industry (start-up companies that grew up in academia), by launching a network of CubeSats built by university teams from 23 different countries around the world to achieve scientific objectives. The key objectives of the mission include facilitating access to space for universities and research centers, performing measurements in the thermosphere, demonstrating new technologies in orbit and promoting space collaboration and science education.

The QB50 network conducts coordinated measurements on a poorly studied and previously inaccessible zone of the lower thermosphere. The project monitors different gaseous molecules and electrical properties to better understand space weather and its long-term trends and relations to climate change. QB50 provides data that enhances atmospheric models and improves understanding of how space weather can disrupt radio communications and GNSS signals. This research contributes to risk assessment of strong solar events that can damage power grids and space assets (i.e., military, commercial and civil satellites).

The project, coordinated by the QB50 Consortium, received funding from the European Union's Seventh Framework Programme for Research and Technological Development. Space agencies are not pursuing a multi-spacecraft network for in-situ measurements in the lower thermosphere because the cost of a network built to industrial standards would be extremely high and not justifiable in view of the limited orbital lifetime. Studying the physics and chemistry of the middle and lower thermosphere can only be realized by using a network of very low-cost satellites, and CubeSats built by universities were the only realistic option.

To accomplish the science mission, 44 highly miniaturized instruments were developed by a consortium of three Universities (UCL-MSSL, the University of Dresden and the University of Oslo) (Fig. 3.7). Thus, QB50 also furthers understanding of how to manufacture, deploy and use small, distributed sensor technologies of the sort that are becoming more common in space.

Fig. 3.7　**Three types of QB50 science sensors (from left to right): the Ion-Neutral Mass Spectrometer (INMS), the Flux-U-Probe Experiment (FIPEX) and the multi-Needle Langmuir Probe (m-NLP). Image Credit: The QB50 Consortium (https://www.qb50.eu/)**

A large portion of the QB50 constellation (28 out of 36 CubeSats) lifted off on April 18th, 2017, from the launch Pad at Cape Canaveral to the ISS and were deployed into space a month later (Fig. 3.8). A second launch was made in June 2017 with the remaining eight CubeSats taking measurements along a polar orbit. Among the 36 CubeSats deployed, 9 were dead on arrival or went silent immediately after launch. For most of the 27 "survivors", commissioning proved to be challenging and only 16 were producing valuable science data on a daily basis. One IOD CubeSat successfully completed its mission within 2 months after launch: InflateSail deployed a dragsail and reentered in the atmosphere. The CubeSats orbited around the Earth, dropping gradually in altitude before completely burning up in the atmosphere, with an estimated lifetime between 1 and 2 years. As of May 2018 (one year after deployment) only 6 CubeSats were still fully operational. During their long descent, the satellites took a large number of measurements using a widely-distributed network of sensors. The last QB50 CubeSats reentered the atmosphere in December 2018, nineteen months after deployment.

QB50 was extremely successful in achieving its educational goals. The QB50 CubeSats were designed, built and operated by a great number of young engineers, supervised by experienced staff at their universities and guided by the QB50 project through reviews and feedback. Those young engineers will leave their universities with valuable hands-on experience. Although the scientific objectives were met with mixed success, this model of international partnership serves as an important pathfinder for future large constellation missions. When backed by adequate resources from national space agencies, such a model could provide enormous potential for science.

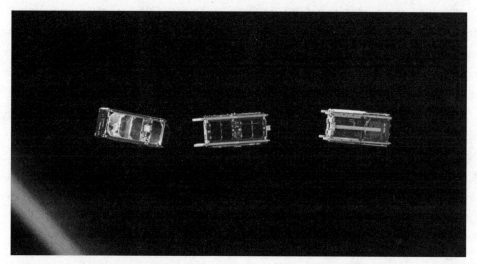

Fig. 3. 8　QB50 satellites after deployment. Image Credit: The QB50 Consortium, von Karman Institute

## 3.5.2　Higher education and sharing lessons learned

CubeSat science missions provide hands-on training opportunities to develop principal investigator leadership, scientific, engineering, and project management skills among both students and early career professionals. Due to the complex nature of the development process that spreads over multiple scientific and engineering domains, teams of students and researchers must be actively involved in the process, potentially over a number of years and take part in all development stages, achieving a level of skill necessary for achieving significant contribution to science.

SmallSat projects are a good example of a pedagogical process known as Project Based Learning (PBL), where science students are actively studying and gaining experience while working on real-world problems. Active learning methods in the fields of science, technology, engineering and mathematics (STEM) show improved learning outcomes in comparison with traditional teaching methodologies.

Lessons can be shared between universities and other organizations in a number of areas.

### 3.5.2.1  Common curriculum development and training methods

The curricular context of CubeSat design activities at universities varies from case to case and many universities do not have a formal CubeSat course curriculum. Instead, the CubeSat projects tend to be integrated as student projects within system engineering or spacecraft design courses (NASEM, 2016). Therefore, there is a clear need to develop a common curriculum where students follow comparable courses in different universities, to facilitate project activities and work at an international level.

A global educational network in academia is addressed by UNISEC[1] (University Spacecraft Engineering Consortium), where worldwide activities of Universities with spacecraft design activities are integrated. Here educational materials related to CanSats and CubeSats are shared, joint workshops and conferences are organized, spacecraft design competitions are organized and standardization efforts to support exchange of subsystems/components are promoted. [2]

### 3.5.2.2  The SpaceMaster case study

The EU Erasmus Mundus "SpaceMaster-Master in Space Science and Engineering" is a joint international MSc program initiated in 2005 and supported by six European Universities. In this integrated program the students can study in three countries, taking advantage of a very broad spectrum for specialization in space science and engineering disciplines. [3] The universities contribute their special expertise to the courses in order to cover the broad interdisciplinary area of spacecraft design and space environment. In particular, system design techniques are emphasized, which are of interest for a broad spectrum of industrial applications well beyond aerospace. The students follow either the more scientific tracks on space physics (with an emphasis on instrumentation and astronomy, or atmospheric and planetary physics) or an

---

[1]  http://www.unisec-global.org.

[2]  http://unisec-europe.eu/standards/bus/.

[3]  http://www.spacemaster.uni-wuerzburg.de, http://spacemaster.eu/.

engineering track emphasizing design of spacecraft and mission realization.

The international dimension of this space education is reflected by study in different locations: first semester in Würzburg (Germany) and second semester in Kiruna (Sweden), second year according to the desired specialization in one of the six partner European Universities. The successful students will receive double diplomas from the two European Universities where most credits were received. The student population is also very international: typically, from about 600 applications 50 students are selected, half of them from Europe, the other half from outside Europe. A specific highlight is the opportunity to participate in small satellite design activities for the MSc thesis. In the UWE program (University Würzburg's Experimental satellite), so far the student-built pico-satellites UWE-1 [1] studying "internet in space" (launched 2005), UWE-2 devoted to attitude determination (launched 2009), and UWE-3 addressing attitude control (launched 2013) have been successfully operated in orbit. The objective of the UWE-program is a step-by-step development of all relevant technologies for formation flying with pico-satellites.

### 3.5.2.3　Sharing lessons learned from SmallSat missions

In the National Academies study (NASEM, 2016) it was established that the failure rate of a university's first CubeSat was typically higher than that of the third or fourth, and that these lessons learnt helped them build better spacecraft. International space conferences (such as COSPAR and IAC), where students and researchers can meet their counterparts from other universities and high-level representatives from space agencies and space industry, provide an opportunity for groups to learn from the experience of others and share lessons learnt. Informal exchanges over lectures or seminars can spark conversations and lead to new initiatives and collaborations.

### 3.5.2.4　Collaborative research requiring complex architectures or constellations

Collaborative university constellations are a low-cost alternative to constellations built by industry to industrial standards. University constellations are also suited to science missions with limited orbital life-time, or having no commercial interest in the industry.

### 3.5.2.5　Sharing resources and standards

International cooperation between universities in the SmallSat field offers significant advantages in cost sharing and thus potential to attract new sponsors to the space science field beyond the classical space agencies. In order to further promote this, electrical interface standards (such as UNISEC-Europe) for a broad application range need to be further developed and extended in order to support exchange of components between scientific institutions. According to the National Academies report (NASEM, 2016), subsystems such as power boards and communication systems standardized to the CubeSat form factor can now be purchased off the shelf. Advances in purchased spacecraft subsystems and common software now permit a science-driven CubeSat mission to focus primarily on developement of the science instrumentation and focus on the science mission. SmallSat projects provide an opportunity to share resources and define standards in the following fields:

• Use of frequencies allocated to Amateur Radio users (VHF, UHF and SBand). University ground stations are capable of communication with more than one spacecraft on the same Amateur Radio frequency bands.

• Shared spectrum and common interface for applying for frequency allocation and coordination.

Adequate spectrum allocation is not only a technical issue but also a regulatory one. A common interface for applying for frequency allocation will facilitate the process for universities and ensure a more effective usage of spectrum. In fact, a single ground station can reuse the same frequency to communicate with many satellites at different time slots.

• Global flexible standard for GS SW and on-board SW based on open source.

• University GS are usually in communications footprint less than 3% of the mission time. For 97% of the time the GS is idle. As an example, for the FIREBIRD mission, only 0.5 percent of the high-rate data was received due to the limitations of the telemetry system. A standard GS and spacecraft SW allows for existing ground stations at universities worldwide to link together, communicate with satellites of other universities, and stream the mission data to research centers via the Internet. The model can be expanded to global networks of university GS providing global coverage for all participating universities and access to a larger amount of data from space at low cost. Many critical operations and science missions would benefit from having uninterrupted coverage allowing for a dramatic increase in mission return.

• Standard low-power transmitters.

These can operate under the power constraints of SmallSats while having the benefit of reducing interference to other satellites, thereby increasing the total throughput from space.

- Standard deployers and shared launches.

Most CubeSats are deployed as secondary payloads on large rockets. CubeSats use launch adapters designed to accommodate them on these launch vehicles. These devices can be used for both dedicated CubeSat launches (usually up to 3U in length) and for shared launches consisting of combinations of smaller-sized CubeSats (e. g. , 1U and 2U). Today, there is still no uniform standard for CubeSat deployers which essentially means that many times the S/C design is dependent on the choice of launch broker. For example, in the ISIS QuadPack CubeSat deployer, access to the S/C is made from the top panel whereas the NanoRacks CubeSat deployer provides access to the S/C from a side panel. Accordingly, the position of the S/C access hatch to charge the batteries or connect to the on-board computer also changes. A standard deployer that can accommodate any CubeSat up to 12U and integrated to any launcher will not only be cost-effective but also increase the variety of orbits available for science CubeSats.

- Centralized systems engineering.

One of the most challenging concepts to teach in aerospace engineering is the interdependent subsystems and systems that make a successful space mission (NASEM, 2016). Proper systems engineering ensures that all likely aspects of a project are considered and integrated into a whole. Unfortunately, unlike the aerospace industry, much of academia does not have a well-established discipline of systems engineering nor a legacy of knowledge and experience in this field. As a result, systems engineering is often the weakest link in the university project and may lead to failure of the scientific mission in space. A project involving several universities requires centralized systems engineering managed by experts who will set a uniform and high standard of implementation for all the partners.

- Centralized management of international projects.

Universities collaborating in an international project should create a joint steering group to reduce duplication in common science and technology areas, targeting resources to the most appropriate partner in each field, creating easier interfaces for investigators, provide more consistency to the integration, testing, and launch efforts and provide a common interface to vendors and launch providers.

### 3.5.2.6  The GENSO case study

GENSO (Global Educational Network for Satellite Operation), supported by educational programs of several space agencies under the lead of ESA 2007, was an early attempt to share ground station resources between universities. Currently most active is UNISEC (University Space Engineering Consortium),[1] where about 100 international universities with spacecraft engineering courses participate. Here not only a global ground station network is supported, but also electrical interface standards[2] are developed (Fig. 3. 9), which were already successfully implemented on several European and Japanese CubeSat missions, and form an excellent basis for exchange of subsystems and components in joint international missions.

More recently, a similar project was started in 2014: SatNOGS[3] is an open source global network of satellite ground stations focused on observing and receiving the signal of satellites, particularly low earth orbit (LEO) CubeSats.

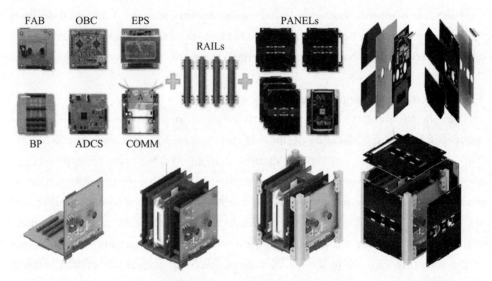

Fig. 3. 9　The flexible composition of a complete satellite through the modular building blocks at subsystem level according to the UNISEC-Europe electrical interface standard, where the harness is replaced by a backplane. Image Credit: Zentrum für Telematik

---

① http://www. unisec-global. org/.

② http://unisec-europe. eu/standards/bus/.

③ https://satnogs. org/.

## 3.5.3  Secondary education

As democratization of space expands, so does the demand for skilled personnel to accomplish simple and complex industrial tasks. Whereas higher scientific education at university and graduate levels is recognized and promoted, a particular effort should be envisaged oriented towards promoting scientific as well as technical careers in secondary level education. A large variety of skilled jobs will be offered in the manufacturing and assembly of SmallSats, not necessarily requiring Msc or PhD level education. "Not only rocket scientists build rockets!"

The secondary education level is well-suited to orient young women and men towards vocational schools and/or technical universities. This means that scientists and industry must define their requirements and, whenever possible, take the time to explain the fascinating field of Space to teachers and students. This would allow schools to set up corresponding curricula and industry to have access to the required workforce. A clear societal benefit of such an approach is the early creation of production jobs providing salaries, taxes and experience.

This approach is implemented, e. g. in Switzerland's National Centre for Competence in Research in Robotics. ① The center provides spin funds that allow scientists or engineers to take work that they have produced in an academic environment and create a spin-off company with it. They are supported in developing their project that has practical applications for the public or companies, thus acquiring the vision and wide range of skills necessary to take something to market. Outside of the academic work, roboticists are required in many organisations such as hospitals, manufacturing plants, environmental services, as well as in space agencies and industry.

One of the problems that educators face today is how to create a teaching environment that provides a meaningful and effective learning experience in the fields of science and technology. Project-based learning (PBL) is also successfully used at the secondary level. PBL is a dynamic classroom approach, where high-school students acquire a deeper knowledge through active research of real-world problems. PBL has been shown to be effective in enhancing both student learning and excitement. CubeSat projects are PBL magnets for science studies. They are using the appeal of

① https://nccr-robotics. ch/kids-teachers-parents/how-to-be-a-roboticist/.

space to the younger generation, attracting them to STEM, and training and preparing them to become the scientists and engineers of tomorrow.

### 3.5.3.1 Encouraging young women to engage in science and technology

Women make up nearly half of the US workforce but only 24% of STEM workers, the US Census Bureau reports. SmallSat programs can be used as platforms to encourage young women to engage in aerospace and STEM. In fact, the Israeli CubeSat program had a very successful program working with state religious schools where boys and girls are separated into single-sex classrooms. R & D teams were formed consisting entirely of female students. Not surprisingly, it turned out that girls not only had excellent R & D capabilities but also leadership and entrepreneurial abilities, and the quality of their tech project was generally better than that of male students of the same age.

### 3.5.3.2 Agency support of teachers and students

In recent years, ESA has been supporting the European Space Education Resource Office (ESERO) project, which envisages the establishment of contact/resource centers which are staffed by education experts and integrated into national educational systems and networks. The centers share inspirational materials that assist teachers and students with the learning process, and supports educational outreach activities that bridge between projects, students and teachers. Several programs exist where high-school students participate in R & D of affordable science experiments that can be flown on various microgravity platforms, such as balloons, or sent to the ISS (Fig. 3.10).

### 3.5.3.3 The Duchifat case study

Duchifat is a CubeSat-based program in the Israeli secondary education system that involves students aged 12-18 years. They start their training as early as the seventh grade with basic science courses. At the ninth grade, students who excel in their studies and show increased motivation, continue to a third year of a more advanced course focused on CubeSat design and become members of a "Satellite and Space Lab" in school. Each team of students is led by an experienced engineer from the aerospace industry and assumes responsibility on one of the satellite's subsystems. The team's task is usually fairly narrow and well-defined, allowing the students to deal with it successfully even though they lack formal engineering education. System engineering, integration and testing issues are also the responsibility of the students but

Fig. 3.10　A trio of CubeSats from Spain, Greece and Israel as seen from the ISS.
Image Credit: NASA

are presented at a later stage (usually the 12th grade) when the students are more experienced and become themselves mentors and leaders to the younger students. This program has already resulted in two CubeSats in space ("Duchifat 1" launched in June 2014, and "Duchifat 2", AKA Hoopoe, launched in May 2017 as part of the QB50 project, both are still fully operational as of May 2018).

Ten additional CubeSats in this series, Duchifat 3 to 12, for ecological applications and space weather monitoring, are now under various stages of development throughout Israel and will be launched by 2020.

Another step forward in creating a new ecosystem that combines academia, industry, government and the education system, is the INDIA/ISRAEL@75 program, a joint venture by India and Israel to develop, build, and launch into space 75 satellites by 2022, celebrating 75 years of Independence in both countries. The satellites will be built by 75 Israeli and Indian high schools and universities to form a constellation that will cover the face of the planet. These rather basic CubeSats (sized between 1U and 3U) will be capable of uploading algorithms from the ground and will serve as a platform for scientific experiments as well as for testing future technologies. The constellation will be controlled and commanded by ground control stations to be set up in schools and universities in both countries. In this novel ecosystem (Academia-

Industry-Education-Governments) the teaching staff will be based primarily on science teachers ( math, physics and computer sciences ) as well as researchers from engineering and exact sciences faculties, but will also include experienced engineers and experts in relevant disciplines from the Israeli and Indian aerospace industries. The staff will guide students in mixed teams of all ages and levels, from high school students to doctoral students. The program is based on the heritage and experience gained in building CubeSats in both countries and is supported by both governments.

## 3.5.4  Fostering international collaboration

There are several other existing frameworks of collaboration between countries. The BIRDS satellite project[①] is a cross-border interdisciplinary project for non-space faring countries supported by Japan (participating countries are Ghana, Mongolia, Nigeria and Bangladesh). During this two-years project students design, develop and operate five units of identical 1U CubeSats. The International Partnership Program (IPP)[②] was launched by the UK in 2015 to deliver a sustainable, economic or societal benefit to undeveloped nations and developing economies.

The current model for selection of large spacecraft that involve international collaborations is not well suited for small satellites. Historical examples such as Solar C and the International Solar Polar Mission (ISPM) illustrate this difficulty due to differences in programmatic frameworks of the involved partners. The process by which QB50 was incepted was a step in the right direction.

COSPAR has a long tradition of Capacity Building Workshops[③] with various partners in order to convey practical knowledge in areas of interest to COSPAR and to build lasting bridges between scientists. This could be developed further into a process equivalent to the decadal surveys but at the international or global level. COSPAR could possibly fill a leading role in such a process.

---

① http://www.birds-project.com.
② https://www.gov.uk/government/collections/international-partner-ship-programme.
③ https://cosparhq.cnes.fr/events/cb-workshops/.

**Finding 3.11** COSPAR as the first and most authoritative international space organization is in a good position to support the international community in the creation and coordination of infrastructure or tools for a global and even deep-space network of small satellites to which anyone can contribute in a well-defined format and interface, thus creating a virtual constellation from all contributors that will by far exceed what the individual parts could do by themselves.

# 4 Recommendations

Based on the findings distributed throughout the text above, we conclude by making five recommendations; one each to the science community, to space industry, to space agencies, to policy makers, and finally, to COSPAR.

## 4.1 Recommendation 1—To the science community

The science community as a whole should acknowledge the usefulness of small satellites and look for opportunities to leverage developments in the small satellite industry. All branches of space science can potentially benefit from the smaller envelope, the associated lower cost, and higher repeat rate. Scientific communities from small countries in particular may benefit from investing their budgets in small satellites.

## 4.2 Recommendation 2—To space industry

Satellite developers should seek out opportunities to partner with individual scientists and universities as well as larger government agencies. This might include data sharing arrangements, selling space on commercial spacecraft for scientific instruments, etc. Currently, publicly available operational data is very valuable for achieving science objectives. Commercial entities should be open to agreements that would continue to make such data available under a free, full, and open data policy for scientific use. Such partnerships can also contribute to workforce development.

## 4.3 Recommendation 3—To space agencies

Large space agencies should adopt procedures and processes that are appropriate to the scale of the project. Agencies should find new ways to provide opportunities for science, applications, and technology demonstrations based on small satellites and with ambitious time to launch. Agencies should additionally take advantage of commercial data or commercial infrastructure for doing science in a manner that preserves open data policies. Finally, space agencies should work together to create long-term roadmaps that outline priorities for future international missions involving small satellites.

## 4.4 Recommendation 4—To policy makers

In order for scientific small satellites to succeed, the scientific community needs support from policy makers to: ① ensure adequate access to spectrum, orbital debris mitigation and remediation options, and affordable launch and other infrastructure services; ② ensure that export control guidelines are easier to understand and interpret, and establish a balance between national security and scientific interests; ③ provide education and guidance on national and international regulations related to access to spectrum, maneuverability, trackability, and end-of-life disposal of small satellites.

## 4.5 Recommendation 5—To COSPAR

COSPAR should facilitate a process whereby International Teams can come together to define science goals and rules for a QB50-like, modular, international small satellite constellation. Through an activity like the International Geophysical Year in 1957-1958 (IGY), participants would agree on the ground rules. Agency or national representatives should be involved from the beginning. The funding would come from the individual participating member states for their individual contributions, or even from private entities or foundations. The role of COSPAR is one of an honest broker, coordinating, not funding. COSPAR should define criteria that must be met by these international teams for proposing.

The results of such an international effort would be valuable for all of the participants, and be more valuable than the individual parts. COSPAR would create a precedent for setting up community science in a very open way. The incentive for participants would be to be part of a worldwide project with access to data of the entire consortium. This recommendation is a means to facilitate progress towards really big ideas such as our four visions for the future or similar ideas.

# 5   Epilogue: Then and now

In the first years of the space age decisions were made quickly and programs started and completed just as quickly: The Soviet Union launched the first satellite on 4 October 1957; NASA was formed less than a year later in July 1958; and Project Apollo started in 1961. In the next year, 1962, the NASA Advisory Council asked the Space Studies Board of the National Academy of Sciences (NAS) to produce a set of high-priority objectives for space science. The first Orbiting Solar Observatory (OSO 1) was launched in March 1962.

The unmanned test flight of the huge Saturn V rocket occurred on 9 November 1967; the first manned Saturn V flight occurred on 11 October 1968; the first flight to the Moon started on 21 December 1968; and then the Moon landing quickly followed on 20 July 1969. During the Apollo Project OSO 3, 4, 5, 6, and Skylab were launched.

Project Apollo was accomplished without e-mail, Excel, PowerPoint or computers with anywhere near the capacity of a low-end smartphone today. Communications occurred by letters, phone, or Fax. To be fair we should recall that the fiscal environment was also very different back then, with cost much less of an issue in the cold war era. Moreover, accountability rules are much stronger today than they were back then.

That was then, but now... The NASA Parker Solar Probe was launched recently (August 2018) and the ESA/NASA Solar Orbiter is in final testing and will follow in 2020. The Solar Probe was recommended by the 2002 Decadal Survey for the Sun and Heliosphere. Before that recommendation, there had been a number of years of project planning. The ESA Orbiter had its origins in 1994. Both of these programs have had at least 20 years of active development, about twice as long as the time to decide to go to the Moon and land a man there.

Both probes approach close to the Sun, so they are complex technical tasks, but it would be hard to argue that they even approach the technical challenges of Project Apollo. Further, the two spacecraft of the Helios Mission in the seventies approached as close to the Sun as Solar Orbiter. Orbiter and Probe are just machines that do not require the oversight of manned mission.

Orbiter and Probe are not isolated examples. The James Webb Telescope was recommended a few years before them and its current earliest launch date is in March 2021, if final testing goes as planned.

Both NASA and ESA have recognized that their science missions take a long time and have developed programs designed to shorten development cycles. NASA has the SMEX and Earth Ventures programs, and ESA has created Class-S missions. These programs operate on a schedule of about 7 years until the first science data are received. The planned schedules also require that the planned rate of development funding is maintained.

A mission's capabilities depend on integrated circuit computers and memory. For several decades the number of components and hence the capabilities of integrated circuits have been doubling every 18 months. This has resulted in a situation where a mission now in orbit is limited by its computers to execute modern software systems that require fast processors and large memory. A few clicks on the web and a few hundred dollars can get you two terabytes of solid state memory, which is much more than is flying on current SMEX missions.

The fact that the current generation of missions are technologically outdated at launch represents an inefficiency in both the usage of funds and human technical and scientific resources. Advances made during the development phase can even make some of the missions science goals obsolete. Understandably, because the missions are so infrequent only those with the lowest perceived technological and scientific risk pass through the long sequence of previews and selection gates.

But things are changing. A new generation of commercial rockets can launch missions at lower cost, thus reducing one of the largest fractions of mission cost. One may therefore expect that in the near future more nations will play a significant role in space science. At present the national governments in China, India, Australia, Korea, and the United Arab Emirates are vigorously developing space science programs to join the four large space agencies, NASA, ESA, Roskosmos, and JAXA/ISAS.

In 1999 two professors, one at California Polytechnic State University and the other at Stanford University, wrote a specification for small satellites used for student experiments in space. The spacecraft was a modular design based on 10 cm cubes-

CubeSats. Their plan was to create projects that would encourage students to become involved in space experiments. CubeSats have evolved over the last decade. They now carry significant scientific payloads. The change has occurred because of the combination of new, lower-cost access to space and light-weight, low-power CPUs in combination with large-capacity memory chips that have been demonstrated to survive in Low Earth Orbit. The CubeSats, and their larger brothers the Nano-Sats and SmallSats, are providing new opportunities for doing science in space both faster and cheaper. This, together with the fact that they can interact in constellations, could create a new era in space science.

The large space agencies are no longer the only players in space operations. In the last years venture capitalists have increasingly recognized that monitoring the Earth from space can yield marketable data. Commercial projects have launched hundreds of CubeSats and SmallSats. The cost of a SmallSat program, while significant, is not too large for Universities to have their own space programs. This is already occurring in Japan, Germany, Israel, Italy, France, the UK, Switzerland, Korea, and others. In 2016, the National Academies published a report on "Achieving Science with CubeSats" (NASEM, 2016). This 2019 COSPAR roadmap reports on the prospects for scientific uses of SmallSats now and in the future.

As stated at the outset, the ultimate destination is a world in which international teams of scientists pursue novel and far-reaching goals. This roadmap provides some possible paths to reach such goals using small satellites. Science missions with masses of tens or a few hundred kilograms instead of tons, development times of a few years instead of decades, and total costs of tens of millions instead of billions may become the norm. The potential of such missions will be amplified further by building constellations of small satellites, thus not only providing multiple observation vantage points but also adding fault tolerance as failure of single network nodes little affect the entire network. A fleet of thousands of networked Earth observation satellites could allow uses and applications of enormous scientific and societal impact. A swarm of small satellites sent to a unique solar system body such as 1P/Halley, each making different observations and built by a different agency, has the potential to outperform any monolithic mission. This is even more apparent for astronomy in space as obviously nothing bigger than the JWST can possibly be launched. And when thinking about reaching any other star within a human lifetime small satellites will have to grow significantly smaller still before such a mission will come within reach even remotely. For all of these and similar visionary goals there are formidable technological challenges to master, but equally importantly, new ways and means of international collaborations between all participating entities—the scientific community in universities and research institutions, space agencies, space industry, policy makers such as governments and international organizations will need to be established, and we hope that COSPAR can play an active and vital role in this process.

# Acknowledgements

This study was supported by the International Space Science Institute (ISSI) with the hosting and funding of two Forum meetings and by COSPAR with travel support to some of the participants.

# 作者信息

| 姓名 | 所属单位 |
| --- | --- |
| Robyn M. Millan | Department of Physics and Astronomy, 6127 Wilder Lab, Dartmouth College, Hanover, NH 03755, USA |
| Rudolf von Steiger | International Space Science Institute, Hallerstrasse 6, CH-3012 Bern, Switzerland<br>Physikalisches Institut, University of Bern, Sidlerstrasse 5, CH-3012 Bern, Switzerland |
| Meir Ariel | Tel Aviv University, The Porter School of Environmental Sciences, P.O. Box 39040, Tel Aviv 6997801, Israel |
| Sergey Bartalev | Space Research Institute (IKI), Russian Academy of Sciences 84/32 Profsoyuznaya Str., Moscow 117997, Russia |
| Maurice Borgeaud | European Space Agency (ESA), ESRIN, Largo Galileo Galilei 1, I-00044 Frascati, Italy |
| Stefano Campagnola<br>Julie C. Castillo-Rogez<br>Graeme Stephens | Jet Propulsion Laboratory, 4800 Oak Grove Drive, Pasadena, CA 91109, USA |
| René Fléron | Technical University of Denmark, Elektrovej Building 327, DK-2800 Kgs. Lyngby, Denmark |
| Volker Gass | Swiss Space Center, EPFL, PPH 339, Station 13, CH-1015 Lausanne, Switzerland |

| 姓名 | 所属单位 |
| --- | --- |
| Anna Gregorio | University of Trieste, Department of Physics, Via A. Valerio 2, I-34127 Trieste, Italy<br>Istituto Nazionale di Fisica Nucleare, Via A. Valerio 2, I-34127 Trieste, Italy<br>Istituto Nazionale di AstroFisica, Osservatorio Astronomico di Trieste, Via Giambattista Tiepolo 11, I-34143 Trieste, Italy |
| David M. Klumpar | Montana State University, Department of Physics, Space Science and Engineering Laboratory, P. O. Box 173840, 264 Barnard Hall, Bozeman, MT 59717-3840, USA |
| Bhavya Lal | IDA Science and Technology Policy Institute, 1701 Pennsylvania Avenue, NW, Suite 500, Washington, DC 20006-5803, USA |
| Malcolm Macdonald | Mechanical & Aerospace Engineering, University of Strathclyde, 75 Montrose Street, Glasgow G1 1XJ, Scotland, UK |
| Jong Uk Park | Space Science Division, Korea Astronomy and Space Science Institute, 776, Daedeok-Daero, Yuseong-Gu, Daejeon 34055, Republic of Korea |
| V. Sambasiva Rao | Centre for Research in Space Science and Technology, PES University, Bengaluru, Karnataka 560085, India |
| Klaus Schilling | Julius-Maximilians-University Würzburg, Am Hubland, D-97074 Würzburg, Germany |
| Alan M. Title | Lockheed Martin Advanced Technology Center, Stanford University, 607 Marion Place, Palo Alto, CA 94301, USA |
| Ji Wu | National Space Science Center, Chinese Academy of Sciences, 1 Nanertiao, Zhongguancun, Haidian District, Beijing 100190, China |

# 缩 略 语

| 缩略语 | 英文全称 | 中文全称 |
|--------|----------|----------|
| 4S | Small Satellites for Space Science | 空间科学小卫星 |
| 5G | Fifth Generation（standard for mobile internet and telephony） | 第 5 代移动通信技术（移动网络和电话标准） |
| AAReST | Autonomous Assembly of a Reconfigurable Space Telescope | 可重构的自主装配空间望远镜 |
| ADCS | Attitude Determination and Control System | 姿态测定与控制系统 |
| AIDA | Asteroid Impact and Deflection Assessment | 小行星撞击和偏转评估 |
| AIM | Aeronomy of Ice in the Mesosphere | 中层大气高空冰探测 |
| AMPTE | Active Magnetic Particle Tracer Explorer | 磁层粒子主动示踪探测器 |
| ARPA | Advanced Research Projects Agency | （美国）先进研究计划局 |
| ARPA-E | Advanced Research Projects Agency-Energy | （美国）能源部先进能源研究计划署 |
| ASI | Agenzia Spaziale Italiana | 意大利航天局 |
| ASTERIA | Arcsecond Space Telescope Enabling Research in Astrophysics | 弧秒空间望远镜天体物理学基础研究卫星 |
| AU | Astronomical Unit | 天文单位 |
| CAS | Chinese Academy of Sciences | 中国科学院 |
| CERN | Conseil Européen pour la Recherche Nucléaire | 欧洲核子研究中心 |

| 缩略语 | 英文全称 | 中文全称 |
|---|---|---|
| CHARA | Center for High Angular Resolution Astronomy | (佐治亚大学)高角分辨率天文中心 |
| CHEOPS | CHaracterising ExOPlanets Satellite | 系外行星表征卫星 |
| CNES | Centre National d'Etudes Spatiales | 法国国家空间研究中心 |
| COBE | Cosmic Background Explorer | 宇宙背景探测器 |
| COSPAR | Committee on Space Research | 国际空间研究委员会 |
| COTS | Commercial Off-The-Shelf | 商用货架产品 |
| CSLI | CubeSat Launch Initiative | 立方星发射计划 |
| CSSWE | Colorado Student Space Weather Experiment | 科罗拉多学生空间天气实验 |
| CYGNSS | Cyclone Global Navigation Satellite System | 飓风全球导航卫星系统 |
| DARPA | Defense Advanced Research Projects Agency | 美国国防部先进研究计划局 |
| DDOR | Delta Differential One-way Ranging | 差分单向测距 |
| DISCUS | Deep Interior Scanning CubeSat | 深空内部扫描立方星 |
| DOD | Department of Defense | 美国国防部 |
| DSCOVR | Deep Space Climate Observatory | 深空气候观测平台卫星 |
| DTUSat | Danmarks Tekniske Universitet Satellite | 丹麦技术大学卫星 |
| EELV | Evolved Expendable Launch Vehicle | 改进型一次性运载火箭 |
| ELaNa | Educational Launch of Nanosatellites | 纳卫星教育发射 |
| ELT | Extremely Large Telescope | 极大望远镜 |
| EM-1 | Exploration Mission 1 | 探索任务1号 |
| ENVISAT | Environmental Satellite | 环境卫星 |
| EO | Earth Observation | 对地观测 |
| EQUULEUS | EQUilibriUm Lunar-Earth point 6U Spacecraft | 地月拉格朗日平动点的6U航天器 |
| ESA | European Space Agency | 欧洲空间局 |
| ESERO | European Space Education Resource Office | 欧洲空间教育资源办公室 |
| ESO | European Southern Observatory | 欧洲南方天文台 |
| ESPA | EELV Secondary Payload Adapter | EELV搭载有效载荷适配器 |
| EU | European Union | 欧盟 |

| 缩略语 | 英文全称 | 中文全称 |
|---|---|---|
| FAST | Fast Auroral Snapshot Explorer | 极光快速拍摄探测器 |
| FCC | Federal Communications Commission | 联邦通信委员会 |
| FDIR | Fault Detection, Identification and Recovery | 故障检测、识别和恢复 |
| FIPEX | Flux-Φ-Probe Experiment | 通量-Φ-探针实验 |
| FIREBIRD | Focused Investigations of Relativistic Electron Burst Intensity, Range, and Dynamics | 相对论电子爆发强度、范围和动力学特性聚焦探测 |
| FLEX | Fluorescent Explorer | 荧光探测器 |
| FP7 | Framework Programme 7 (of the EU) | 欧盟第七框架计划 |
| GALEX | Galaxy Evolution Explorer | 星系演化探测卫星 |
| GENSO | Global Educational Network for Satellite Operations | 全球卫星运控教育网络 |
| GEO | Geosynchronous Equatorial Orbit | 地球同步轨道 |
| GNSS | Global Navigation Satellite Systems | 全球导航卫星系统 |
| GOLD | Global-scale Observations of the Limb and Disk mission | 边缘与盘面全球尺度观测任务 |
| GPS | Global Positioning System | 全球定位系统 |
| GSSW | Ground System SoftWare | 地面系统软件 |
| GSTP | General Support Technology Programme | 一般支持技术计划 |
| GTO | Geostationary Transfer Orbit | 地球同步转移轨道 |
| GTOSat | Geostationary Transfer Orbit Satellite | 地球同步转移轨道卫星 |
| HERMES | High Energy Rapid Modular Ensemble of Satellites | 高能快速模块化组合卫星 |
| HG | High Gain | 高增益 |
| I & T | Innovation and Technology | 创新和技术 |
| IAC | International Astronautical Congress | 国际宇航大会 |
| IARPA | Intelligence Advanced Research Projects Activity | 美国情报先进研究计划局 |
| IBEX | Interstellar Boundary Explorer | 星际边界探测器 |
| ICE | International Cometary Explorer | 国际彗星探测器 |

| 缩略语 | 英文全称 | 中文全称 |
|---|---|---|
| ICON | Ionospheric Connection Explorer | 电离层连接探测器 |
| IDA | Institute for Defense Analyses | 美国防务分析研究所 |
| IGRF | International Geomagnetic Reference Field | 国际地磁参考场 |
| IGY | International Geophysical Year | 国际地球物理年 |
| IIST | Indian Institute of Space Science and Technology | 印度空间科学与技术研究院 |
| IKAROS | Interplanetary Kite-craft Accelerated by Radiation Of the Sun | "伊卡洛斯"太阳帆飞船 |
| INMS | Ion-Neutral Mass Spectrometer | 离子与中性质谱仪 |
| INSPIRE | Interplanetary Nano-Spacecraft Pathfinder inRelevant Environment | 相关环境中的行星际纳卫星探路者 |
| IOD | In-Orbit Demonstration | 在轨验证 |
| IPP | International Partnership Program | 国际伙伴关系项目 |
| IRIS | InfraRed Imaging Surveyor | 界面区成像光谱仪 |
| ISA | Israeli Space Agency | 以色列航天局 |
| ISAS | Institute of Space and Astronautical Science | 空间和宇航科学研究所 |
| ISIS | Innovative Solutions in Space company | 空间创新解决方案公司 |
| ISO | InterStellar Object | 星际天体 |
| ISPM | International Solar Polar Mission | 国际太阳极轨道任务 |
| ISRO | Indian Space Research Organisation | 印度空间研究组织 |
| ISS | International Space Station | 国际空间站 |
| ITAR | International Traffic in Arms Regulations | 国际武器贸易条例 |
| ITU | International Telecommunications Union | 国际电联 |
| JAXA | Japan Aerospace Exploration Agency | 日本宇宙航空研究开发机构 |
| JPL | Jet Propulsion Laboratory | 喷气推进实验室 |
| KRIS | Ka-band Interferometry Swarm | Ka 波段干涉测量集群 |
| KSAT | Kongsberg Satellite Services | 康斯伯格卫星服务公司 |
| LASP | Laboratory for Atmospheric and Space Physics | 大气和空间物理实验室 |

| 缩略语 | 英文全称 | 中文全称 |
| --- | --- | --- |
| LEO | Low Earth Orbit | 近地轨道 |
| LHC | Large Hadron Collider | 大型强子对撞机 |
| LIGO | Laser Interferometer Gravitational-Wave Observatory | 激光干涉引力波天文台 |
| LISA | Laser Interferometer Space Antenna | 激光干涉空间天线 |
| LPO | Long Period Object | 长周期天体 |
| LSST | Large Synoptic Survey Telescope | 大型综合巡天望远镜 |
| MarCO | Mars Cube One | 火星立方1号 |
| M-ARGO | Miniaturised Asteroid Remote Geophysical Observer | 微型小行星远程地球物理观测器 |
| Microscope | A microsatellite to challenge the universality of free fall | "显微镜"卫星,即挑战自由落体定律微卫星 |
| MIDEX | Medium Explorer (line of NASA satellites) | 中型探索者计划 |
| MIRO | Microwave Instrument for the Rosetta Orbiter | "罗塞塔"轨道器微波设备 |
| MIT | Massachusetts Institute of Technology | 麻省理工学院 |
| MMS | Magnetospheric Multiscale Mission | 磁层多尺度任务 |
| m-NLP | multi-Needle Langmuir Probe | 多针朗缪尔探头 |
| NANOBED | Nanosatellite Applications aNd Operations Bench for Engineering and Demonstration | 纳卫星应用和操作基准工程验证 |
| NASA | National Aeronautics and Space Administration | 美国国家航空航天局 |
| NASEM | National Academies of Sciences, Engineering, and Medicine | 美国国家科学工程和医学院 |
| NCLE | Netherlands Chinese Low Frequency Explorer | 荷兰-中国联合低频探测器 |
| NOAA | National Oceanic and Atmospheric Administration | 美国国家大气海洋局 |
| NSF | National Science Foundation | 美国国家科学基金会 |

| 缩略语 | 英文全称 | 中文全称 |
|---|---|---|
| NTIA | National Telecommunications and Information Administration | 国家电信和信息局 |
| NuSTAR | Nuclear Spectroscopic Telescope Array | 核光谱望远镜阵列 |
| OCSD | Optical Communications and Sensor Demonstration | 光通信与传感器演示验证 |
| OLFAR | Orbital Low Frequency ARray | 轨道低频阵列 |
| OSCAR-1 | Orbiting Satellite Carrying Amateur Radio 1 | 业余无线电爱好者轨道卫星-1 |
| Pan-STARRS | Panoramic Survey Telescope And Rapid Response System | 泛星计划,即全景巡天望远镜和快速反应系统 |
| PBL | Project Based Learning | 项目式学习 |
| PROBA | Project for On-Board Autonomy | 星上自主项目 |
| PROCYON | Proximate Object Close flyby with Optical Navigation | 光学导航临近天体近距离飞越卫星 |
| PSLV | Polar Satellite Launch Vehicle | (印度)极轨卫星运载火箭 |
| QB50 | EU project,a network of 50 small satellites in space | 欧盟 50 颗小卫星组网项目 |
| R & D | Research and Development | 研究与开发 |
| REPT | The Relativistic Electron-Proton Telescope | 相对论电子质子望远镜 |
| RHESSI | Ramaty High Energy Solar Spectroscopic Imager | 拉马第高能太阳光谱成像仪 |
| SAMPEX | Solar Anomalous and Magnetospheric Particle Explorer | 太阳异常和磁层粒子探测器 |
| SAOCOM | Satélite Argentino de Observación COn Microondas | 阿根廷微波观测卫星 |
| SAR | Synthetic Aperture Radar | 合成孔径雷达 |
| SatNOGS | Satellite Networked Open Ground Station | 卫星网络开放地面站项目 |
| SEU | Single Event Upset | 单粒子翻转事件 |
| SHALOM | Spaceborne Hyperspectral Applicative Land and Ocean Mission | 星载高光谱陆地和海洋应用任务 |
| SIM | Space Interferometry Mission | 空间干涉测量任务 |

| 缩略语 | 英文全称 | 中文全称 |
|---|---|---|
| SLS | Space Launch System | 空间发射系统 |
| SMEX | Small Explorer (line of NASA satellites) | 小型探索者计划 |
| SMILE | Solar wind Magnetosphere Ionosphere Link Explorer | 太阳风-磁层相互作用全景成像卫星 |
| SPARC | Star-Planet Activity Research CubeSat | 恒星-行星活动研究立方星 |
| STEM | Science，Technology，Engineering and Mathematics | 科学、技术、工程和数学 |
| SWAS | Submillimeter Wave Astronomy Satellite | 亚毫米波天文卫星 |
| TESS | Transiting Exoplanet Survey Satellite | 凌日系外行星勘测卫星 |
| THEMIS | Time History of Events and Macroscale Interactions during Substorms | 亚暴期间的事件演化过程及其大尺度相互作用任务 |
| TIM | Telematics International Mission | 国际远程信息处理任务 |
| TOM | Telematics earth Observation Mission | 远程地球观测任务 |
| TRACE | Transition Region and Coronal Explorer | 过渡区和日冕探测器 |
| TRL | Technology Readiness Level | 技术成熟度 |
| TROPICS | Time-Resolved Observations of Precipitation structure and storm Intensity with a Constellation of Smallsats | "热带"卫星群任务，即基于小卫星星座的降水结构和风暴强度时间分辨观测 |
| UHF | Ultra High Frequency (300~3000 MHz) | 超高频 |
| ULA | United Launch Alliance | 联合发射联盟 |
| UNEX | University Explorer (line of NASA satellites) | 大学探索者计划 |
| UNISEC | University Space Engineering Consortium | 大学空间工程联盟 |
| UV | Ultraviolet | 紫外 |
| UWE | Universität Würzburg Experimental satellit | 伍兹堡大学实验卫星 |
| VCLS | Venture Class Launch Services | 风险级发射服务 |
| VENμS | Vegetation and Environment monitoring on a New Micro-Satellite | 植被和环境监测新型微卫星 |
| VHF | Very High Frequency (30~300 MHz) | 甚高频 |
| VLTI | Very Large Telescope Interferometer | 超大望远镜干涉仪 |
| XML | Extensible Markup Language | 可扩展标记语言 |

# 参 考 文 献

ANDERSON J, BARNWELL N, CARRASQUILLA M, et al., 2018. Sub-nanosecond ground-to-space clock synchronization for nanosatellites using pulsed optical links. Adv. Space Res. 62, 3475-3490. https://doi.org/10.1016/ j. asr. 2017.06.032.

ANDERSON B J, KAZUE T, TOTH B A, 2000. Sensing global Birkeland currents with iridium engineering magnetometer data. Geophys. Res. Lett. 27 (24), 4045-4048.

ANGELOPOULOS V, PANETTA P V (Eds.), 1998. Science closure and enabling technologies for constellation class missions. University of California Publication, Berkeley CA.

ATWATER H A, DAVOYAN A R, ILIC O, et al., 2018. Materials challenges for the starshot lightsail. Nat. Mater. 17, 861-867. https://doi.org/ 10.1038/s41563-018-0075-8.

BAMBACH P, DELLER J, VILENIUS E, et al., 2018. DISCUS-The deep interior scanning CubeSat mission to a rubble pile near-Earth asteroid. Adv. Space Res. 62, 3357-3368. https://doi.org/10.1016/j.asr.2018.06.016.

BANDYOPADHYAY S, FOUST R, SUBRAMANIAN G P, et al., 2016. Review of formation flying and constellation missions using nanosatellites. J. Spacecr. Rock. 53 (3), 567-578. https://doi.org/10.2514/1.A33291.

BASTIDA VIRGILI B, KRAG H, 2015. Small satellites and the future space debris environment. Proceedings of the 30th ISTS Conference, 4-10 July 2015, Kobe,

Japan. https://archive. ists. or. jp/upload_pdf/2015-r-11. pdf.

BENNIS W, BIEDERMAN P W, 1997. Organizing genius: the Secrets of Creative Collaboration. Basic Books, New York.

BERERA A, 2017. Space dust collisions as a planetary escape mechanism. Astrobiology 17 (12), 1274-1282. https://doi. org/10.1089/ ast. 2017.1662.

BOTEZ D, GARROD T, MAWST L J, 2015. High CW wallplug efficiency 1. 5 micron-emitting diode lasers. In: 2015 IEEE Photonics Conference, Reston VA, 4-8 October 2015, pp. 551-552. https://doi. org/10.1109/ IPCon. 2015. 7323726.

BRONIATOWSKI D A, JORDAN N C, LONG A M, et al. , 2005. Balancing the needs for space research and national security in the ITAR. In: Proceedings of the AIAA Conference Space 2005, 30 August-1 September 2005, Long Beach CA, paper 2005-6800. https://doi. org/10.2514/6. 2005-6800.

CAHOY K, 2015. CubeSats in astronomy and astrophysics. Presentation at the SSB CubeSta Symposium, Irvine, CA, 2 September 2015. http://sites. nationalacademies. org/cs/groups/ssbsite/documents/webpage/ssb_167819. pdf.

CAPPELLETTI Ch, BATTISTINI S, GRAZIANI F, 2018. Small launch platforms for micro-satellites. Adv. Space Res. 62, 3298-3304. https://doi. org/ 10.1016/j. asr. 2018. 05. 044.

CASTELVECCHI D, 2018. Chinese satellite launch kicks off ambitious mission to Moon's far side. Nature 557, 478-479. https://doi. org/10. 1038/ d41586-018-05231-9.

CATAPULT, 2017. Small Satellite Market Intelligence. Satellite Applications Catapult Ltd. , Harwell, UK. https://sa. catapult. org. uk/services/market-reports/ small-sats-market-intel/.

CAUSA F, RENGA A, GRASSI M, 2018. Robustfilter setting in GPS-based relative positioning of small-satellite LEO formations. Adv. Space Res. 62, 3369-3382. https://doi. org/10.1016/j. asr. 2018. 03. 020.

COLAGROSSI A, LAVAGNA M, 2018. Fully magnetic attitude control subsystem for picosat platforms. Adv. Space Res. 62, 3383-3397. https://doi. org/10.1016/ j. asr. 2017. 10. 022.

DALE D C, WHITCOMB G P, 1994. Small satellite missions in the context of the ESA scientific programme. ESA Bulletin Nr. 80, 7-14.

ESPER J, NEECK S, SLAVIN J, et al. , 2003. Nano/micro satellite constellations for earth and space science. Acta Astronaut. 52, 785-791. https://doi. org/10. 1016/ S0094-5765(03)00054-7.

FENNELL J F, SPENCE H E, MOORE T E, et al. , 2000. Magnetospheric constellation missions. In: Harris, R. (Ed.), Proceedings of the Cluster-II Workshop: Multiscale/Multipoint Plasma Measurements, vol. 449. European Space Agency (ESA/ESTEC), ESA-SP, Noord-wijk, pp. 235-242.

FLéRON R W, 2017. Will CubeSats introduce a Moore's law to space science missions? In: Graziani, Filippo (Ed.), Advances in Astronautical Sciences, Vol. 163, Fourth IAA Conference on Dynamics and Control of Space Systems, University Satellite Missions and CubeSat Workshop, pp. 677-694, ISBN 978-0-87703-647-0.

FLéRON R W, 2018. Evolution of spacecraft masses before and after the introduction of CubeSats. Presentation at the 4S Symposium, Sorrento, 28 May-1 June, 2018, paper 2_230-333. https://atpi. eventsair. com/QuickEventWebsitePortal/4s2018/4s/ExtraContent/Con-tentSubPage? page = 1&subPage = 1.

FOTI G, GOMMENGINGER Ch, SROKOSZ M, 2017. First spaceborne GNSS-reflectometry observations of hurricanes from the UK Tech-DemoSat-1 mission. Geophys. Res. Lett. 44(24), 12358-12366. https:// doi. org/10. 1002/2017GL076166.

GAIDOS E, WILLIAMS J, KRAUS A, 2017. Origin of interstellar object a/ 2017 U1 in a nearby young stellar association? Res. Notes AAS 1 (1), 13. https://doi. org/10. 3847/2515-5172/aa9851.

GOVINDARAJAN V, 2016. The three box solution. Harvard Business Review Press, Boston MA.

HALT T, WIEGER A, BOENSCH N, et al., 2019. Smallsats by the Numbers 2019. Bryce Space and Technology, DC Metro Chicago London. https://brycetech. com/reports. html.

HOANG T, LAZARIAN A, BURKHART B, et al., 2017. Astrophys. J. 837 (5), 1-16. https://doi. org/10. 3847/1538-4357/aa5da6.

HULOT G, EYMIN C, LANGLAIS B, et al., 2002. Small-Scale Structure of the Geodynamo Inferred from Ørsted and Magsat Satellite Data. Nature 416, 620-623.

IANNOTTA B, 2019. Mass Producer. Aerospace America. https://aerospaceamerica. aiaa. org/departments/mass-producer/.

KLESH A T, BAKER J D, BELLARDO J, et al., 2013. INSPIRE: Interplanetary NanoSpacecraft Pathfinder in Relevant Environment. In: AIAA SPACE 2013 Conference, September 10-12, 2013, San Diego, CA, paper 2013-5323. https://doi. org/10. 2514/6. 2013-5323.

LAL B, DE LA ROSA BLANCO E, BEHRENS J R, et al., 2017. Global trends in small satellites IDA paper P-8638. IDA Science & Technology Policy Institute, Washington, D.C..

LUBIN P, 2016. A roadmap to interstellar flight. J. Br. Interplanet. Soc. (JBIS) 69, 40-72, arXiv: 1604.01356.

LUPO L, ALBANESE C, BETTINELLI D, et al. , 2018. Lighthouse: a spacebased mission concept for the surveillance of geosynchronous space debris from low Earth orbit. Adv. Space Res. 62, 3305-3317. https://doi. org/10. 1016/j. asr. 2018. 03.005.

MATNEY M, VAVRIN A, MANIS A, 2017. Effects of CubeSat deployments in low-earth orbit. Proceedings of the 7th European Conference on Space Debris, 18-21 April 2017, Darmstadt, Germany. https://ntrs. nasa. gov/search. jsp? R = 20170003837.

MCNUTT R, WIMMER-SCHWEINGRUBER R F, KRIMIGIS S M, et al. , 2018. Interstellar probe: the first step of a thousand miles. 42nd COSPAR Scientific Assembly, Pasadena CA, July 14-22, 2018, Scientific Assembly Abstracts 42, Abstract PIR. 1-0001-18, 2366-2366.

MEECH K J, WERYK R, MICHELI M, et al. , 2017. A brief visit from a red and extremely elongated interstellar asteroid. Nature 552 (7685), 378-381. https:// doi. org/10. 1038/nature25020.

MISTICHELLI F, 2016. NOAA current and future satellite networks. Presentation to the Commercial Smallsat Spectrum Management (CSSMA) Meeting, Washington, D.C. , 16 September 2016. https:// www. cssma. space.

NASA, 2015. Small spacecraft technology state of the art. NASA Ames Research Center, Moffett Field, CA, NASA/TP-2015-216648/Rev1.

NASEM, 2016. Achieving science with CubeSats: thinking inside the box. The National Academies Press, Washington, DC (doi: 10. 17226/ 23503). https:// www. nap. edu/catalog/23503/achieving-science-with-cubesats-thinking-inside-the-box.

National Research Council, 2010. An enabling foundation for NASA's space and earth Science missions. The National Academies Press, Washington, DC, 10. 17226/12822.

NOCA M, JORDAN F, STEINER N, et al. , 2009. Lessons learned from the first Swiss Pico-satellite: SwissCube. Proceedings of the AIAA/USU Conference on Small Satellites, Technical Session XII: The Next Generation, Paper SSC09-XII-9. https://digitalcommons. usu. edu/smallsat/2009/ all2009/84/ .

PAPITASHVILI V O, CHRISTIANSEN F, NEUBERT T, 2002. A new model offield-aligned currents derived from high-precision satellite magnetic field data. Geophys. Res. Lett. 29 (14), 1683. https://doi. org/10. 1029/ 2001GL014207.

PARHAM J B, BEUKELAERS V, LEUNG L, et al., 2019. Leveraging commercial cubesat constellations for auroral science: a case study. J. Geophys. Res. Space Phys. 124, 3487-3500. https://doi.org/10.1029/2018JA025966

PEñA V, HOWIESON S V, LAL B, et al., 2017. Early stage research and technology at U. S. Federal Government Agencies IDA Document D-8481. IDA Science & Technology Policy Institute, Washington, D.C..

PEREZ F, MODENINI D, VáSQUEZ A, et al., 2018. DustCube, a nanosatellite mission to binary asteroid 65803 Didymos as part of the ESA AIM mission. Adv. Space Res. 62, 3335-3356. https://doi.org/10.1016/j.asr.2018.06.019.

PIETRZAK A, WOELZ M, HUELSEWEDE R, et al., 2015. Heading to 1 kW levels with laser bars of high-efficiency and emission wavelength around 880 nm and 940 nm. In: Zediker, M. S. (Ed.), High-Power Diode Laser Technology and Applications XIII, Proc. SPIE 9348, 93480E. https://doi.org/10.1117/12.2078642.

RAAB W, BRANDUARDI-RAYMONT G, WANG C, et al., 2016. In: Space Telescopes and Instrumentation 2016: Ultraviolet to Gamma Ray, Proc. SPIE 9905, p. 990502. https://doi.org/10.1117/12.2231984.

ROTTEVEEL J, CHAMOT B, BENTUM M, et al., 2017. A road map for low frequency radio astronomy in lunar orbit using CubeSats. iCubeSat 2017 Workshop, Cambridge, UK, 30-31 May, 2017, paper 2017.A.2.2.

RUF C S, CHEW C, LANG T, et al., 2018. A new paradigm in earth environmental monitoring with the CYGNSS small satellite constellation. Sci. Rep. 8. https://doi.org/10.1038/s41598-018-27127-4. Article #8782.

SANTONI F, SEITZER P, CARDONA T, et al., 2018. Optical tracking and orbit determination performance of self-illuminated small spacecraft: LEDSAT (LED-based SATellite). Advances in Space Research 62, 3318-3334. https://doi.org/10.1016/j.asr.2018.08.018.

SAUNDERS C, LOBB D, SWEETING M N, et al., 2017. Building large telescopes in orbit using small satellites. Acta Astronaut. 141, 183-195. https://doi.org/10.1016/j.actaastro.2017.09.022.

SEN A, 2014. Transformative innovation: what "totally radical" and "island-bridge" mean for NOAA research dissertation. The George Washington University.

SEN A, 2015. Organized to transform: what NOAA can learn from DARPA, ARPA-E and ATP. STPI Brownbag Presentation, January 28, 2015. https://prezi.com/3tpusuohmpny/organized-to-transform-what-noaa-can-learn-from-darpa-arpa-e-and-atp/? utm_campaign = share&utm_medium = copy.

SHAWHAN S D, 1990. Role of small satellite missions in magnetospheric research. In: Hultqvist, B., Fälthammar, C.-G. (Eds.), Magnetospheric Physics. Springer, Boston, MA, pp. 103-114. https://doi.org/10.1007/ 978-1-4615-7376-0_8.

SPENCE H, MOORE T, KLIMAS A, et al., 2004. Global dynamics of the structured magnetotail: updated synopsis of the report of the NASA Science and Technology Definition Team for the Magnetospheric Constellation Mission. http://www.phy6. org/MagCON.pdf.

SHKOLNIK E L, ARDILA D, BARMAN T, et al., 2018. Monitoring the high-energy radiation environment of exoplanets around low-mass stars with SPARCS (Star-Planet Activity Research CubeSat). Presentation at the American Astronomical Society Meeting #231, Washington, DC, January 8-12, 2018 (Abstract 228.04).

STUPL J, MASON J, MARSHALL W, et al., 2012. LightForce: orbital collision avoidance using ground-based laser induced photon pressure. In: Phipps, C. (Ed.), International Symposium on High Power Laser Ablation, AIP Conference Proceedings, Vol. 1464, pp. 481-491. https://doi.org/ 10.1063/1.4739902.

SWEETING M N, 2018. Modern small satellites-changing the economics of space. Proc. IEEE 106, 343-361. https://doi.org/10.1109/ JPROC.2018.2806218.

THIRY N, VASILE M, 2014. Recent advances in laser ablation modelling for asteroid detection methods. In: Taylor, E. W., Cardimona, D. A. (Eds.), Nanophotonics and Macrophotonics for Space Environments VIII, Proc. SPIE 9226, p. 922608. https://doi.org/10.1117/12.2060810.

TOUBOUL P, MéTRIS G, RODRIGUES M, et al., 2017. MICROSCOPE mission: first results of a space test of the equivalence principle. Phys. Rev. Lett. 119, 231101. https://doi.org/10.1103/PhysRevLett. 119.231101.

TSUDA Y, MORI O, FUNASE R, et al., 2011. Flight status of IKAROS deep space solar sail demonstrator. Acta Astronaut. 69 (9-10), 833-840. https:// doi.org/10. 1016/j.actaastro.2011.06.005.

TUMLINSON J, 2014. CYGNSS: lessons learned from a class D mission. Presentation at NASA Workshop on EEE Parts for Small Missions. NASA GSFC, September 10-11, 2014. https://nepp.nasa.gov/work-shops/eeesmallmissions/talks/11%20-%20THUR/0900%20-%20NEPP%20EEE%20Parts%20for%20Small%20Missions%20CYGNSS%20Presentation.pdf.

VIQUERAT A, SCHENK M, LAPPAS V, et al., 2015. Functional and qualification testing of the InflateSail technology demonstrator. In: 2nd AIAA Spacecraft Structures Conference, Kissimmee FL, Paper 2015-1627. https://doi.org/10.

2514/6. 2015-1627.

WALKER R, 2018. A Hitchhikers' guide to the solar system: future missions of deep space nano -spacecraft. Presentation at the 4S Symposium, Sorrento, 28 May-1 June, 2018, paper 2_29-79. https://atpi. event-sair. com/QuickEventWebsitePortal/4s2018/4s/ExtraContent/Con-tentSubPage? page = 1&subPage = 1.

WALL M, 2018. NASA's next planet hunter will launch a new era of exoplanet research, space. com, March 28, 2018. https://www. space. com/40128-nasa-tess-mission-new-era-exoplanet-research. html.

ZURBUCHEN T H, VON STEIGER R, BARTALEV S, et al. , 2016. Performing high-quality science on CubeSats. Space Res. Today 196, 10-30.

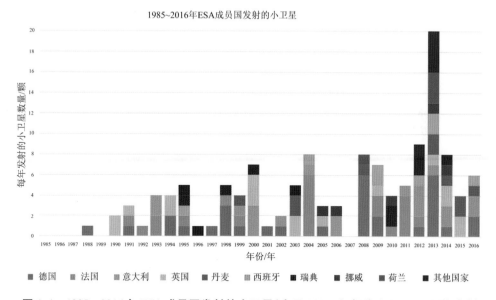

图 1.1　1985～2016 年 ESA 成员国发射的小卫星(小于 200 kg)；相比之下，ESA 同期发射
了 6 颗小卫星[数据收集由拉尔等人(Lal et al. ,2017)完成，数据由拉尔提供]

图 1.2　1957~2017 年各国每年发射的小卫星(小于 200 kg)数量[本图转载自 Lal 等人(2017)的报告"全球小卫星趋势"图 E-2,并已经过防务分析研究所同意]

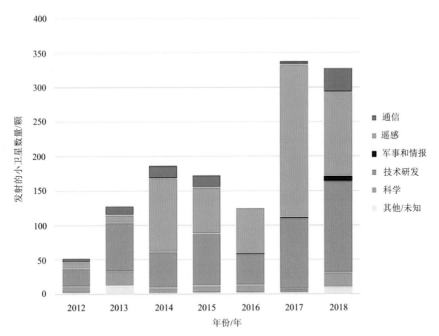

图 1.3　各种用途的小卫星数量[图片来源:Bryce Space and Technology (https://brycetech.com/reports.html)]

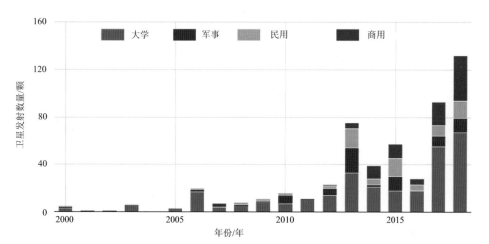

图 1.4　2000～2018 年每年立方星的发射数量,根据负责设计/建造/运行卫星的机构类
　　　　型划分(大学立方星、军事立方星、民用立方星、商业立方星)。本图不包含自
　　　　2014 年开始由 Planet 和 Spire 项目研制的 450 多颗商业星座立方星[图表由 M.
　　　　Swartwout 制作,使用数据截止到 2018 年底(数据来源:https://sites. google.
　　　　com/a/slu. edu/swartwout/home/cubesat-database)]

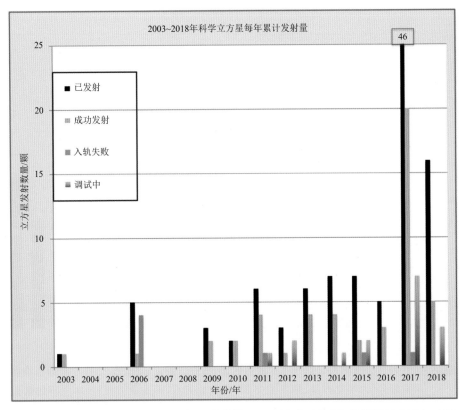

图 1.5 科学立方星:2003～2018 年累计发射了 107 颗;仅 2017 年就发射了 46 颗。107
颗立方星中的 7 颗因为发射失败而未能入轨。图中标注 4 个类别:已发射立方
星、成功立方星、入轨失败、立方星调试中。另外 2 个类别未体现在图中,分别
为到达预定位置时已失灵(此类共计 21 颗)和立方星状态未知(此类共计 14
颗)(数据来源:Swartwout, https://sites. google. com/a/slu. edu/swartwout/
home/cubesat–database)

图 1.6　NASA"热带"(TROPICS)卫星任务概况[图片来源：TROPICS 团队,麻省理工学院林肯实验室(TROPICS 是"利用小卫星星座进行降水结构和风暴强度的时间分辨观测"的英文缩写),卫星网站链接为 https://tropics. ll. mit. edu/cms // Mission-Overview]

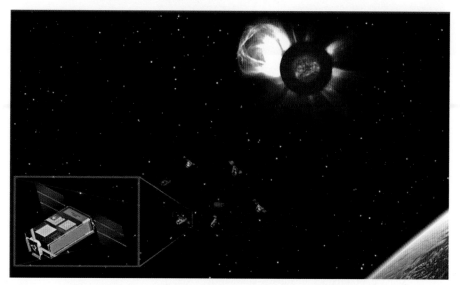

图 1.7　目前,一些空间干涉测量的任务概念正在论证中,其中包括本图所示的 NASA"日出"任务概念(SunRISE mission concept),该任务将采用 6U 立方星组成的小星座来测量太阳射电暴(solar radio bursts)(图片来源："日出"任务概念研究团队)

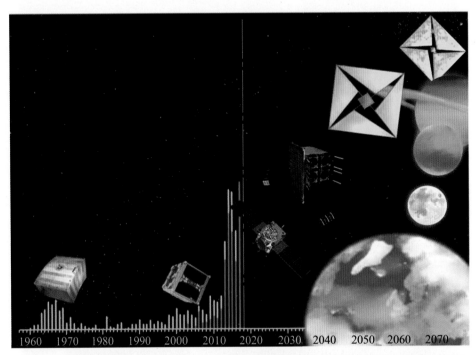

图 1.8　浅色条状图显示了从 1957 年到目前立方星的演变(基于 Swartwout 的数据)。深
　　　　色表示立方星的数量。随着"火星立方任务"完成了火星之旅、SLS-EM-1 任务
　　　　和 Hera 任务即将发射并计划对迪迪莫斯双小行星系统进行探测,"搭载"的发
　　　　射模式已经超出了近地轨道。上述 3 项卫星任务都采用了立方星模式。未来
　　　　的卫星任务可能要求航天器的质量更低,因此需要新的航天器概念,如由深空
　　　　探测任务分离舱部署出来的立方星集群或者激光推进的超小型芯片卫星
　　　　[Katrine Grønlund 艺术假想图:Oscar-1 卫星艺术图获得授权使用:http://www.
　　　　arrl. org/space-communication;DTUsat-1 卫星图像由 DTUsat-1 卫星团队提供;
　　　　Hera 卫星任务图片来源:https://www. esa. int/our_activities/space_engineering
　　　　_technology/asteroid_impact_mission/asteroid_impact_deflection_assessment_
　　　　mission;国际空间站四星释放器立方星图像获得授权使用:https://www.
　　　　isispace. nl/product/quadpack-cubesat-deployer/]

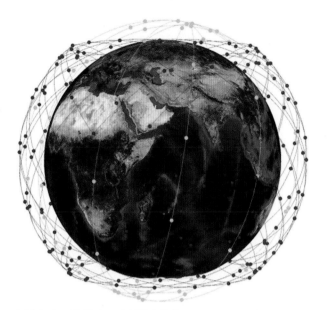

图 2.1　大型商业卫星星座示意图[图片来源：Telesat(加拿大电信卫星公司)]

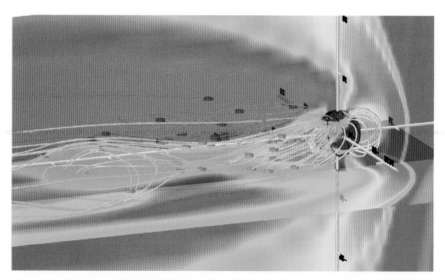

图 2.2　在 NASA 的"磁层星座"(MagCon)任务概念中，36 个航天器叠加在磁层流
体动力学(MHD)模拟的地球磁层上[转自 Spence 等人(2004)]

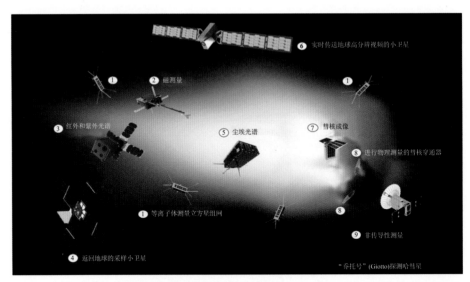

图 2.3　利用分离的星座来探测再次回归的哈雷彗星(2061 年)。本图展示了各种类型的探测手段,来完全揭秘这颗不寻常的彗星(它不属于两类经典的彗星:木星家族和奥尔特云)(图片来源:ESA)

图 2.4　可重构空间望远镜自动组装(AAReST)任务概念:直径为 10 cm 的圆形镜面安装在一组立方星上[图片来源:加州理工学院 AAReST 团队 (http://www.pellegrino.caltech.edu/ aarest1/)]

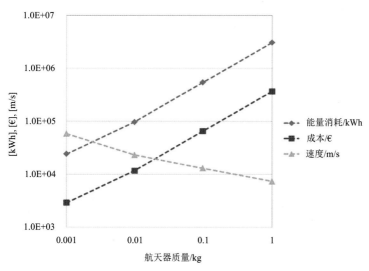

图 2.5　航天器质量与激光阵列能量消耗、成本和最终速度之间的关系图

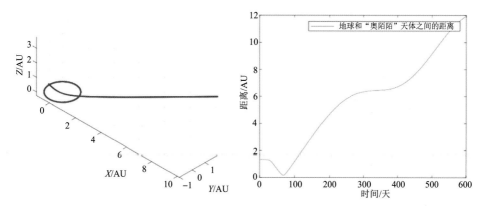

图 2.6　左图:"奥陌陌"('Oumuamua)天体经过地球轨道时的轨道。右图:地球和"奥陌陌"天体之间的总体距离(本图所用位置数据取自 JPL Horizons 数据库和网页:https://ssd.jpl.nasa.gov/horizons.cgi。"奥陌陌"天体是达到最近距离后才被发现的)

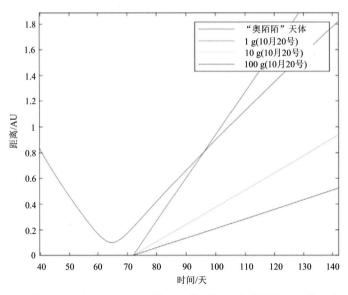

图 2.7 蓝线代表了 2017 年 10 月 20 日"奥陌陌"天体与地球位置之间的距离。本图所示的模拟仿真也是从这个日期航天器所处的太空位置开始加速。红色、黄色和紫色三条线分别代表三种航天器质量,都使用 10 MW 激光阵列进行加速。可见只有质量为 1 g 的航天器才能获得能够追上"奥陌陌"天体的速度

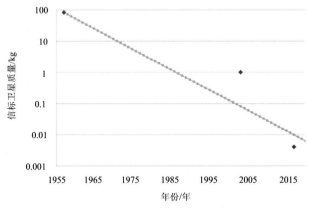

图 2.8 Sputnik(1957)、Cute-1(2003)和 Sprite(2017)等信标卫星的质量演化。灰色趋势线表示质量减半周期为 55 个月

航天器的质量/kg

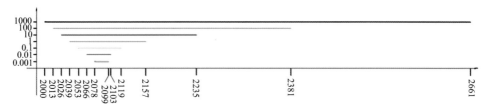

图 2.9　航天器质量与发射、到达时间(目的地是半人马座 α 比邻星)的函数关系。假设航
　　　　天器质量的演化遵循航天器版本的摩尔定律。这里的航天器质量减半时间被设定
　　　　为 4 年(Fléron,2017 年)[航天器速度数据来源于卢宾(Lubin,2016)]

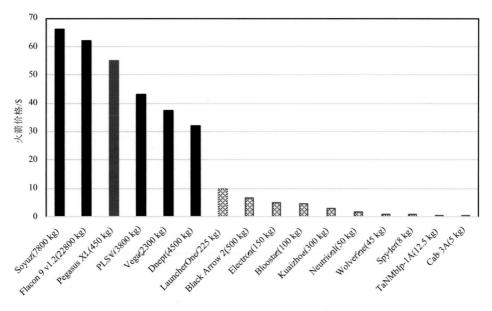

图 3.1　火箭价格与最大发射能力对比(按成本排序)。黑色柱状图代表大型火箭,红色代
　　　　表小型火箭,阴影代表研制中的火箭[本图转载自拉尔等人的文章(Lal et al.,
　　　　2017),并经防务分析研究所许可]

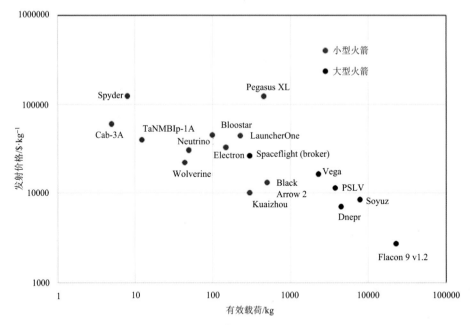

**图 3.2　小型和大型火箭的每千克发射价格[本图转载自拉尔等人的文章(Lal et al.,2017),并经防务分析研究所许可]**

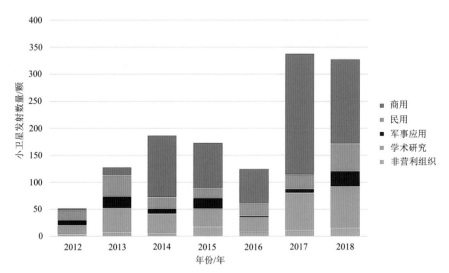

**图 3.3　各领域发射的小卫星数目[图片来源:布莱斯空间和技术公司(https://brycetech.com/reports.html)]**

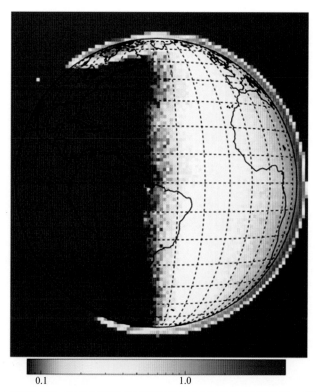

0.1　　　　　　　　　　1.0

图 3.4　NASA 的"边缘与盘面全球尺度观测"(GOLD)任务捕捉到的第一张地球高层大气
　　　　层紫外原子氧释放图像(135.6 nm 级)(图片来源:NASA/大气和空间物理实验室
　　　　GOLD 任务科学团队)

图 3.5　连接飞行硬件的纳卫星应用和操作基准工程验证(NANOBED)项目的核心硬
　　　　件装置,位于斯特拉斯克莱德(Strathclyde)大学实验室[图片来源:马尔科姆
　　　　·麦克唐纳(Malcolm Macdonald)]

图 3.6　3 颗"国际远程信息处理任务"(TIM)卫星通过摄影测量方法聚焦
同一目标区域的 3D 图像[图片来源：Zentrum fur Telematik
(https://www.telematik‑zentrum.de/)]

图 3.7　欧盟 QB50 项目的三种类型科学传感器(从左到右)：离子与中性质谱仪(INMS)、通
量-Φ-探针实验(FIPEX)和多针朗缪尔探头(m-NLP)[图片来源：欧盟 QB50 项目科
学小组(https://www.qb50.eu/)]

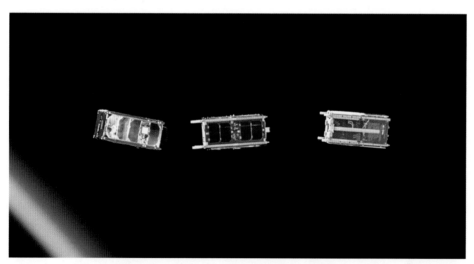

图 3.8　部署后的欧盟 QB50 卫星(图片来源:欧盟 QB50 卫星科学小组,冯·卡门研究所)

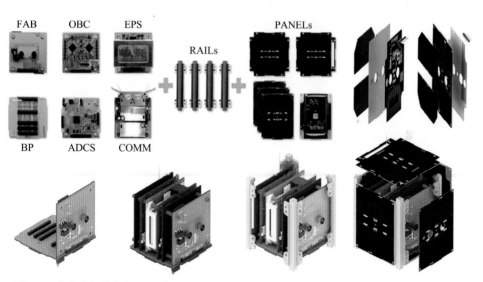

图 3.9　根据"大学空间工程联盟-欧洲"电气接口标准,通过子系统级别的模块化组件,可
灵活组装完整卫星,其中线束由背板代替(图片来源:Zentrum fur Telematik)

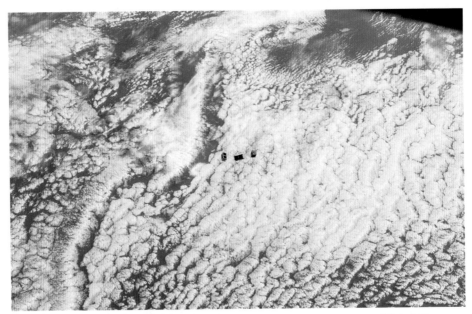

图 3.10　从国际空间站上观看 3 颗分别来自西班牙、希腊和以色列的立方星
（图片来源：美国国家航空航天局）